Multilayered Low Temperature Cofired Ceramics (LTCC) Technology

Multilayered Low Temperature Cofired Ceramics (LTCC) Technology

Yoshihiko Imanaka

Fujitsu Laboratories, Ltd.
Japan

Library of Congress Cataloging-in-Publication Data

A C.I.P. Catalogue record for this book is available
from the Library of Congress.

ISBN 978-1-4419-3576-2   Printed on acid-free paper.
e-ISBN 978-0-387-23314-7
©2010 Springer Science+Business Media, Inc.

Printed in the United States of America.

9 8 7 6 5 4 3 2 1

springeronline.com

This book is dedicated to ALL I love.

Contents

List of Figures

Chapter 1

Chapter 2

Chapter 3

Chapter 4

Chapter 5

Chapter 6

Chapter 7

Chapter 8

Chapter 9

Chapter 10

List of Tables

Chapter 1

Chapter 2

Chapter 3

Chapter 4

Chapter 5

Chapter 9

Chapter 10

Preface

In recent years, Low Temperature Cofired Ceramics (LTCC) have become an attractive technology for electronic components and substrates that are compact, light, and offer high-speed and functionality for portable electronic devices such as the cellular phones, personal digital assistants (PDA) and personal computers used for wireless voice and data communication in rapidly expanding mobile network systems. For their wiring, these LTCCs use metals such as Cu, Ag, and Au with little conductor loss and low electrical resistance at high frequencies, while the ceramics used in LTCCs have lower dielectric loss than organic materials. This makes LTCCs especially suitable for the high frequency circuits required for high-speed data communications.

During the late 1980s, U.S. and Japanese manufacturers of computers and ceramic materials conducted extensive research and development of LTCC technology that is now crucial to present day and future communications technologies. At that time Fujitsu and IBM America produced a large, multi-layer ceramic substrate (meeting Fujitsu's specifications of 254 × 254 mm with 60 layers) with a copper wire pattern for use in mainframe computers. I was involved in developing and producing this substrate and from the standpoint of manufacturing, the level of technology employed then far exceeds that of current LTCC products. The substrate was produced using very precise control of a host of manufacturing parameters.

However, no book was published concerning the high level of LTCC technology of that time, and I was very concerned that this technology would be forgotten. So in order for this technology to survive into the future, I made a systematic compilation of LTCC technology to serve primarily as a benchmark. This book gives an account from the engineering perspective of the technology development at Fujitsu for the mainframe computer substrate mentioned above. In addition, I present some basic scientific material needed to understand this technology.

In Japan we have an old Chinese saying *"onkochishin"* which means studying the old in order to find new knowledge and techniques. Old things are remade into new things. Ideas of a previous age are redeveloped. I believe this applies as much to technology as to the world of music and fashion. I hope that this book will contribute to efforts to combine elements of earlier LTCC technology with something new, in order to create entirely different technologies.

I myself am an engineer involved in research and development of LTCC technology and I am working on the development of new technologies. These technologies are not complete; they are work in progress. I hope to revise and further expand the material in this first book to offer an even better resource.

July, 20th, 2004
Atsugi, Japan

Yoshihiko Imanaka

Acknowledgments

In pursuing research and development and carrying out experiments at Fujitsu Laboratories for the multilayered LTCC with copper wiring used as a circuit board in Fujitsu's mainframe computer, I had many technical discussions concerning the various problems arising at the production plant, and went through many cycles of trial and error with Mr. Kenichiro Abe, Mr. Yoshikazu Abe and Mr. Tsuyoshi Sakai who were working on development, and I would like to express my profound gratitude to them and my other colleagues at Fujitsu.

I am deeply grateful to Dr. Koichi Niwa and Dr. Nobuo Kamehara, general manager of the Materials & Environmental Engineering Laboratories for their sustained guidance concerning the development of electronic ceramics.

Dr. Michael R. Notis of the Department of Materials Science and Engineering at Lehigh University, my Master's degree advisor, and the late Dr. Kenji Morinaga of Kyushu University, my Doctor's degree advisor, gave me much help and encouragement in my ceramic material research, and taught me the fascination of materials research and development. I am very grateful to them.

I would like to thank Mr. Gregory T. Franklin, Senior Publishing Editor of Springer for giving me this opportunity to publish this book, and to Ms. Carol Day, Editorial Assistant, for her assistance with the administrative procedures required for publishing this book, including reading the English proofs. I am also grateful to Mr. Rod Walters for his extensive assistance in checking the English text, and to Mr. Susumu Zenyoji for creating the cover design.

I would like express my gratitude to my father Eizaburo and my mother Atsuko who are still active in the fields of culture and the arts.

Finally, to my wife Sachiko who continuously gave her devoted support for my health and well-being, and to my son Koki whose innocent smile and delightful dance inspired me, I offer my thanks.

Chapter 1

Introduction

With the current the explosive growth of mobile phones, communication technology for transmitting text and image data wirelessly using mobile phones as the terminal device is continually making rapid progress. At the same time, a wide variety of applications are being found for broadband and high frequency technologies. Using waves of 800 MHz, 1.5 GHz, and 2 GHz, mobile phones are moving to increasingly higher frequencies. Bluetooth (2.45 GHz) for wireless LANs, ETC (5.3 GHz) and the like are finding commercial applications, and 2 GHz or higher is increasingly being used [1, 2]. In the 10 GHz or higher quasi-millimeter waveband too, plans are underway to introduce WLL (Wireless Local Loop, 20 to 30 GHz) and in-car radar (50 to 140 GHz, and 76 GHz shows promise) [3]. In order to realize further progress in high frequency wireless communication technologies of this kind, system solution development in tandem with hardware technology development will likely play the important role of bringing greater multi-functionality, higher performance, and sub-miniaturization to mobile terminal devices. For example, mobile terminal devices are being equipped with multiple functions such as Bluetooth, GPS, and wireless LAN, so in order to curb the resulting increases in circuit size, it is desirable to build the various high frequency functions and passive components into the substrate itself rather than mounting them on its surface. In addition, to enable high speed data communication, it is anticipated that electronic components and substrates meeting the requirement for high frequencies and broad bandwidth, and with little loss at high frequencies can be achieved [4, 5].

As it is comparatively easy to combine Low Temperature Cofired Ceramics (LTCC) with materials that have different characteristics, it is possible to integrate and build the different types of components into the ceramic. Furthermore, while it is possible to incorporate low loss metal into LTCCs as a conductor, ceramic has low dielectric loss at high frequencies making it effective for achieving low loss performance, compared with other materials such as resin and the like. In addition, its thermal expansion coefficient compared with resin materials and other ceramic materials is low, and it has the merit of excellent connection reliability for high density packaging of LSI components. For these reasons, LTCCs are regarded as a promising future technology for the integration of components and substrates for high frequency applications.

1.1   Brief historical review

The origin of multilayer ceramic substrate technology is said to lie in developments at RCA Corporation in the late 1950s, and the bases of current process technologies (green sheet fabrication technology, via forming technology, and multilayer laminate technology using the doctor blade method) were discovered at this time [6-8]. Thereafter, progress was made using these technologies with IBM taking the lead, and the circuit board (board size: 9 cm^2, with 33 layers, and 100 flip chip bonded LSI components) for IBM's mainframe computer commercialized in the early 1980s was the inheritance [9, 10, 11]. Since this multilayer board was Cofired at the high temperature of 1,600°C with the alumina insulating material and conductor material (Mo, W, Mo-Mn), it is called High Temperature Cofired Ceramic (HTCC) to distinguish it from the Low Temperature Cofired Ceramics (LTCC) developed later. From the middle of the 1980s, efforts to increase the speed of mainframe computers accelerated, and as the key to increasing computer performance, further improvements were made to multilayer ceramic substrates for high density mounting applications. By using finer wiring in order to increase wiring density in circuit boards for high density mounting, the electrical resistance of the wiring increases, and conspicuous attenuation of the signal occurs. Therefore it is necessary to use materials with low electrical resistance (Cu, Au or the like) for the wiring. In addition, with the flip chip method of connecting bare LSI components directly, poor connection of the interconnects may result if the thermal expansion of the board is not close to that of the silicon components (3.5 × 10^{-6}/°C), therefore an insulating material with low thermal expansion (ceramic) is desirable. Furthermore, to achieve high speed transmission of signals, it is necessary to ensure that the ceramic has a low dielectric constant. By the early 1990s, many Japanese and American electronics and ceramics manufacturers had developed multilayer boards (LTCC) that met these requirements[12, 13]. Among them, Fujitsu and IBM were the first to succeed with commercial applications of multilayer substrates using copper wiring material and low dielectric constant ceramics[14, 15]. From the latter half of the 1990s to the present, the focus of applications has shifted to high frequency wireless for the electronic components, modules and so on used in mobile communication devices, primarily mobile phones. For the multilayer circuit board, the low thermal expansion of ceramics was its biggest merit for the purposes of high density mounting of LSI components. However, for high frequency communications applications, its low transmission loss is its key feature, and the low dielectric loss of ceramic gives it an advantage over other materials.

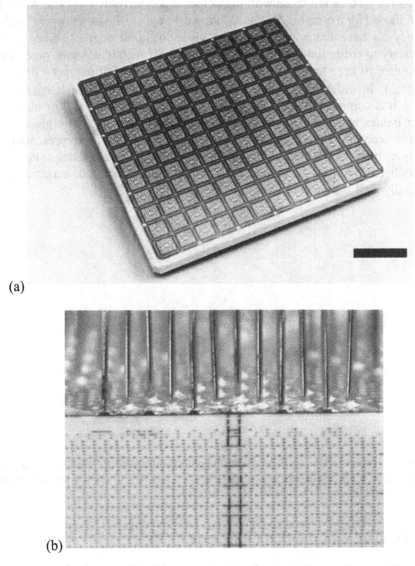

(a)

(b)

Figure 1.1 (a) Multilayered ceramic circuit board for mainframe computers produced by Fujitsu (Size: 245 × 245 mm, 52 layers) (Bar = 50 mm) (b) Cross-sectional view of the circuit board with Cu in internal wiring (via diameter: 80 μm, conducting line width: 80 μm, line spacing: 100 μm, dielectric material thickness per layer: 200 μm).

1.2 Typical material

As its name suggests, LTCC is ceramic cofired with metal wiring at low temperature, and its constituent materials are metal and ceramic. The typical metals for LTCCs are those with high electric conductivity (Ag, Cu, Au and their alloys [for example, Ag-Pd, Ag-Pt, Au-Pt etc.]), and as shown in Table 1, they all have a low melting point close to 1,000 degrees. Since it is necessary to cofire the ceramic material with these metals, extreme precision is required to keep temperatures below the melting point of the metal (900 to 1,000°C). In order to ensure high sintered density with low temperature firing, it is common to add amorphous glass, crystallized glass, low melting point oxides and so on to the system to enhance sintering. The glass and ceramic composite such as that shown in Figure 1.2 is a representative ceramic material. Besides this type, crystallized glass, composites of crystallized glass and ceramic, and liquid phase sintered ceramic are generally well known types.

Figure 1.2   Glass/Alumina composite with 20 vol% alumina content. Dielectric constant of the composite is 5.6, thermal expansion coefficient is $3.5 \times 10^{-6}/°C$, thermal conductivity is 2.4 W/mK, and flexural strength is 200 MPa [Bar = 5 μm].

1.3 Typical manufacturing process

The basic manufacturing process for multilayer ceramic substrates is shown in Figure 1.3 [16]. First, the ceramic powder and organic binder are mixed to make a milky slurry. The slurry is cast into tape using the doctor blade method, to obtain a raw ceramic sheet (green sheet) that before firing, is flexible like paper. Vias for conduction between layers and wiring patterns are screen printed on the green sheet using conductive paste. Many layers of these printed green sheets are arranged in layers, and heat and pressure is applied to laminate them (the organic resin in the green sheets acts as glue for bonding the layers during lamination). By firing the conductor metal and ceramic together while driving off organic binder in them, a multilayer ceramic substrate can be obtained. The most important point to bear in mind in the manufacturing process is controlling variation in the dimensional precision and material quality of the finished product, and process conditions must be set so that the micro and macro structures of the work in progress are homogenous at every process step.

Table 1.1   Typical material combination of LTCC and HTCC.

	Ceramics		Conductor	
	Material	Firing temperature (°C)	Material	Melting point (°C)
LTCC	· Glass/Ceramic composite · Crystallized glass · Crystallized glass/Ceramic composite · Liquid-phase sintered ceramics	900 to 1,000	Cu	1,083
			Au	1,063
			Ag	960
			Ag-Pd	960 to 1,555
			Ag-Pt	960 to 1,186
HTCC	· Alumina ceramics	1,600 to 1,800	Mo	2,610
			W	3,410
			Mo-Mn	1,246 to 1,500

Furthermore, the technique of laminating and cofiring more than two types of ceramic sheet with different dielectric characteristics, and the process of forming a resistor by cofiring are also well known [17].

1.4 Typical product types

Figure 1.4 shows a block diagram of a dual band mobile phone with GPS. To take transmission as an example, the analog signal of your voice when you speak is first converted into a digital signal by an AD converter. Next, using a mixer, an operation is performed to mix the signal with a high frequency component, and its overall frequency is raised. Then noise is eliminated using a SAW filter, and additionally, the strength of the signal is increased through amplification and it is transmitted from an antenna.

LTCCs are used in the individual components that are used for performing these processes – the coupler (which controls PA output gain on the PA output side), and balun (a device which transforms balanced and unbalanced impedance). Similarly, LTCCs are used for the boards that form the module circuits of the sort shown in the diagram (the antenna switch module, front end module, and power amp module). In addition, the SAW filter package is also made using LTCC. As this suggests, there are many LTCC products built into the circuits of high frequency mobile terminals. It is expected that the parts shown with a dotted line in the diagram will in future also use LTCC modules. Product types can be divided into three types as in Table 1.2 – discrete components, boards and packages, and modules – and all types are currently used as products for wireless communication in commercial applications.

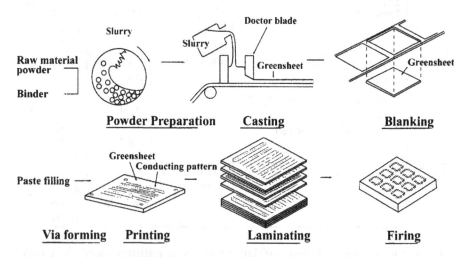

Figure 1.3   Typical multilayer ceramic substrate manufacturing process.

1.5 Characteristics of LTCC

Compared with printed resin boards, LTCCs are superior from the three aspects of high frequency characteristics, thermal stability, and their capacity for integrating passive components. They are also well suited to integrated substrates and electronic components for high frequency applications.

1.5.1 High frequency characteristics

High frequency transmission loss ($1/Q$) is expressed as the relationship between dielectric loss ($1/Q_d$) and conductor loss ($1/Q_c$). As shown in Figure

1.5 (a), dielectric loss is the loss of charge accumulated between the conductor and ground in the transmission line, causing the frequency to get higher, so that the leakage of current increases and the flow of the current in the conductor is impeded. Dielectric loss is generally shown using the following formula

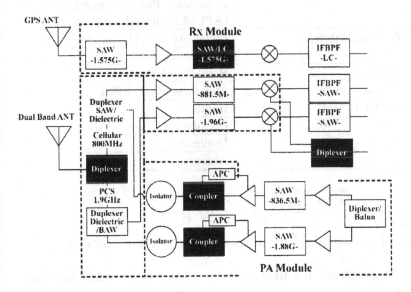

Figure 1.4   Block diagram of a dual band mobile phone (CDMA, PCS) with GPS.

$$1/Q_d = \frac{20\pi \cdot \log e}{\lambda_g}\tan\delta = 2.73 \cdot \frac{f}{c}\sqrt{\varepsilon_r} \cdot \tan\delta \ \ (dB/m)$$

λg: wave length, f: frequency, c: light velocity, ε_r: dielectric constant, tan δ: dielectric loss tangent

However, conductor loss depends on the resistance (surface resistance) of the conductor. As the frequency increases, there is a tendency for the current to concentrate in the surface parts of the conductor. The part where the current flows is known as skin depth (the depth where current density falls to $1/e = 0.37$ of its value at the surface), and it decreases in inverse proportion to the square root of the frequency. Surface resistance R_s is determined by skin depth d and conductor conductivity σ as in the formula below. It is inversely proportional to the square root of conductor conductivity, and increases proportional to the square root of the frequency.

Table 1.2   Classification of typical LTCC products.

Type	Product
Module	Front-end Module
	Rx Module
	Automatic Power Control, (APC)/Coupler Module
Package/Substrate	Power Amp (PA) Module
	Saw device Package
Surface Mount Device (SMD)	Band Pass Filter (BPF)
	Low Pass Filter (LPF)
	Balun
	Coupler
	Duplexer
	Antenna

$$Rs = \frac{1}{d \cdot \sigma} = \sqrt{\frac{\pi \cdot f \cdot \mu_0}{\sigma}} = \sqrt{\pi \cdot f \cdot \mu_0 \cdot \rho}$$

f: frequency, μ_0: permeability of vacuum, ρ: conductor resistance

Figure 1.5 (b) shows the frequency dependence of dielectric loss and conductor loss. It shows the result using a model with conductor thickness of 30 μm, width 8 mil (203.2 μm), tan δ 0.02, ε_r 3.5, and characteristic impedance of 50 Ω. At frequencies lower than 1 GHz, conductor loss is more dominant than dielectric loss as far as signal attenuation is concerned. However, above 1 GHz, the impact of dielectric loss becomes all the more conspicuous with increases in frequency.

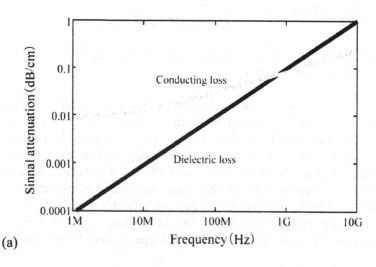

(a)

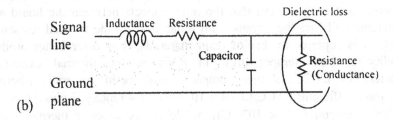

Figure 1.5   The frequency dependence of dielectric loss and conductor loss (a) and dielectric loss in a circuit (b).

Table 1.3   Comparison between the dielectric characteristics of ceramic material and resin material.

Material		Dielectric constant (@ 2 GHz)	tan δ [Q value] (@ 2 GHz)
Ceramics	Soda-lime glass	6.8	0.01 [95]
	Borosilicate glass	4.5	0.006 [150]
	Silica glass	3.8	0.00016 [6,000]
	Alumina	9.0	0.0003 [3,000]
	LTCC (Alumina/Borosilicate glass)	5 - 8	0.005 - 0.0016 [200 - 600]
Organic material	Epoxy	3.1	0.03 [30]
	FR4 (Epoxy + EGlass 60 wt%)	4.3	0.015 [65]
	Polyimide	3.7	0.0037 [270]
	Teflon (PTFE)	2.0	0.0005 [2,000]

As shown above, in order to reduce the loss at high frequencies over 1 GHz, it is beneficial to use dielectric material with low tan δ. As shown in Table 1.3, in general tan δ is lower for ceramic materials compared with resin materials. Furthermore, tan δ of LTCCs is 1/3 lower than that of FR4, the material used for printed resin boards, and compared with resin printed circuit boards, it is more suitable for high frequency applications.

1.5.2 Thermal stability (low thermal expansion, good thermal resistance)

Circuit boards and packages undergo heat stress during assembly processes such as in solder reflow when LSI components and other electronic parts are mounted on them, and when reliability tests are performed before product shipment, so it is a concern that the interconnects between the board and components will lose their connection reliability. As the thermal resistance of LTCCs is superior to that of resin materials, they offer better product reliability at high temperatures. In addition, the thermal expansion coefficient of LTCC is low compared with resin materials [thermal expansion coefficient: LTCC ($3 - 4 \times 10^{-6}/°C$), FR4 (epoxy/E-glass) ($16 - 18 \times 10^{-6}/°C$), silicon ($3.5 \times 10^{-6}/°C$)], while it has superior thermal shock resistance giving it better thermal stability including connection reliability, than resin printed circuit boards. Representative reliability test specs are as follows.

- Pressure cooker test – temperature: 120°C, humidity: 100 RH%, testing time: 50 h
- Heat exposure test – temperature: 150°C, testing time: 1,000 h
- Temperature cycle test: -65°C to RT to 125°C, test cycles: 100

1.5.3 Integration of passive components

If the many passive components currently mounted on substrates can be built into the board, the length of the wiring used to connect these passive components can be shortened, so that improved characteristics can be achieved such as making the board itself smaller, and reducing the parasitic inductance. Additionally, as the space freed up on the newly formed substrate can be used to mount other components and the like, the board can be given more functions.

With this method of manufacture, it is easy to introduce different kinds of materials in sheet form, and it is possible to form layered structures composed of appropriate dissimilar materials in order to make the LTCC manifest a variety of passive functions. In order to embed passive functions in the substrate, green sheets of different materials are stacked and integrated in the LTCC, and as they are inserted in layers, there is considerable freedom in design. In addition, because special proper material is used for the desired functions, improved functioning can be expected. For example, as shown in Figure 1.6 (a), a passive function can be integrated in an LTCC by combining a material with a low dielectric constant of around 5 for the signal wiring to enable high speed transmission, a medium dielectric constant material of around 15 used as a filter, and a material with a high dielectric constant of 1,000 or more for eliminating signal noise, providing a power source to lowered potential and so on [18, 19]. With resin printed circuit boards on the other hand, it is not possible to introduce material with a high dielectric constant in sheets inside the substrate due to the constraints of the method of manufacture and process temperatures. For this reason, a method has been proposed whereby, after components are mounted on the board as shown in (b), a top sheet of insulating material is formed and a sheet material consisting of epoxy resin mixed with high dielectric constant ceramic particles (dielectric constant of about 50) suitable for low temperature processes is inserted [20, 21, 22, 23]. However, since the types and characteristics of the materials and the locations for embedding components are limited, the resulting miniaturization arising from integration of components is not so great as with LTCC.

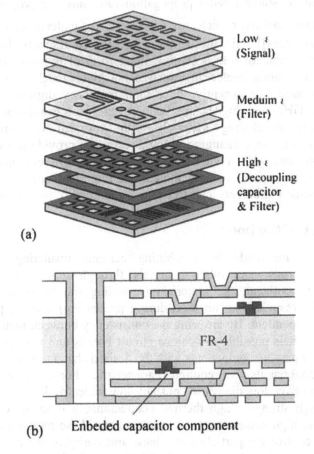

(a)

Low ε
(Signal)

Meduim ε
(Filter)

High ε
(Decoupling
capacitor
& Filter)

(b)   Enbeded capacitor component

Figure 1.6   Passive components embedded in substrates
[LTCC (a), and printed resin board (b)].

1.6 Trends in materials developed by relevant companies

Table 1.4 and 1.5 show the material specifications of the dielectric ceramics for LTCC produced by the relevant companies at the end of the 80s [24, 25,

26, 27, 28, 29] and in 2000 [30]. Items in the table for which the data is unknown are shown blank. As Cu, Au, Ag, or their alloys are used inside the substrate as conductor material, the firing temperature of any of the ceramics is 900 to 1,000°C. As explained in 1.1, in the later 80s LTCCs were developed as circuit boards for mainframe computers with an operating frequency of about 300 MHz – high performance for that time. With the goal of high density mounting of bare LSI chips made of silicon on substrates to enable high speed transmission, ceramic materials were developed that had low dielectric constants (signal propagation rate must be proportional to $C/\sqrt{\varepsilon}$ [C: light velocity, ε: dielectric constant of the material in the vicinity of transmission lines]) and with thermal expansion coefficients close to that of silicon. As conductor loss is the dominant form of transmission loss at these frequency bands, dielectric loss was not given much attention as a development issue. But recently, since high frequency digital and analog signals of 1 GHz and higher are being used for transmission, all the companies are developing materials with low tan δ. Furthermore, developments targets are changing in the direction of providing a variety of dielectric constants with a view to combining the different materials to achieve different passive functions. In addition, moves are apace to make the glass and ceramic ingredients lead-free in consideration of the environment.

1.7 Subject of the book

Since LTCCs are made by combining ceramic insulating materials, conductor materials, and other materials through numerous processes culminating in cofiring, the materials (including the materials such as organic binder resin and so on used during processing) and the processes used are interdependent. By ensuring the consistency between materials and processes first, it is possible to achieve circuit boards and various types of high frequency passive components with the desired characteristics.

Each material has its own required characteristics. For example, ceramics are required to have low dielectric constant, low dielectric loss, low thermal expansion, high strength, high thermal conductance and so on. Conductor metals provide high conductivity. In order to meet these requirements, it is necessary to control the particle size, shape and purity of the raw material powder, as well as the composition of the material. Furthermore, in order to complete all the processes from shaping, printing, laminating through firing ceramics and conductors that have good characteristics and combining them, it is especially important to give the process conditions close scrutiny. However, during processing, if there is no physical or chemical consistency between the different materials, it will be necessary to return to the previous process conditions. On occasion, it may also be necessary to change the composition of the materials themselves.

Table 1.4   Material specifications of LTCCs of the relevant companies
(1985 – 90) [Ref. 24, 25, 26, 27, 28, 29].

LTCC Suppliers	Products (composition)	Dielectric constant (ε)	Resistivity ($\Omega \cdot cm$)	Thermal expansion co-efficient (ppm/°C)	Thermal conduct-ivity (w/m $\cdot$ K)	Flexural strength (MPa)
Asahi glass	Al_2O_3 35 wt% + Forserite 25 wt% + BSG 40 wt%	7.4	>10^{14}	5.9	4.2	235
Kyocera	BSG + SiO_2 + Al_2O_3 + Cordierite	5.0	>10^{14}	4.0	2	190
	Crystallized glass + Al_2O_3	6.2	>10^{14}	4.2	3	210
Dupont	Al_2O_3 + $CaZrO_3$ + Glass	8.0	>10^{12}	7.9	4.5	200
Sumitomo metal ceramics	$(CaO-Al_2O_3-SiO_2-B_2O_3)$ glass 60 wt% + Al_2O_3 40 wt%	7.7	>10^{14}	5.5	2.5	196
NEC	(PbO-BSG) glass 45 wt% + Al_2O_3 55 wt%	7.8	>10^{14}	4.2	4.2	300
Noritake	Al_2O_3 + Forsterite + glass	7.4	5×10^{16}	7.6	8.4	140
Hitachi	(BaO-Al_2O_3-BSG) + Al_2O_3 + $ZrSiO_4$	7.0	10^{13}	5.5	1.7	200
Fujitsu	Al_2O_3 50 wt% + BSG 50 wt%	5.6	>10^{14}	4.0	4.0	200
Matsushita	(PbO-BSG) 45 wt% + Al_2O_3 55 wt%	7.4	>10^{12}	6.0	3.0	260
IBM	Cordierite crystallized glass	5.0	-	-	3.0	210
NGK	ZnO-MgO-Al_2O_3-SiO_2 (Cordierite system)	5.0	5×10^{15}	3.0	3.0	200
Taiyo-yuden	Al_2O_3-CaO-SiO_2-MgO-B_2O_3	7.0	>10^{14}	4.8	8.4	250
Toshiba	$BaSnB_2O_6$	8.5	2×10^{15}	5.4	5.4	200
Murata	BaO-Al_2O_3-SiO_2	6.1	>10^{14}	8.0	2.0	200
Reference	Si	-	10×10^{-6}	3.5	170	-

As one example, consider the interfacial phenomenon between ceramic and copper wiring. If the material or process conditions of the ceramic and copper are inappropriate, various macro and micro flaws occur. For example, in the firing process, minute pores are formed at the interface. Possible causes of the formation of the pores are (1) mismatch of the firing and shrinkage behaviors of the conductor and ceramic, (2) insufficient adherence between the conductor and ceramic in the laminating process, (3) gasification of undecomposed resin residue powder present at the interface, and (4) gas formed through chemical reaction between the conductor and ceramic. In the case of cause (1), it is not sufficient only to optimize the firing process conditions, and in some cases, it may be necessary to revise the base powder of each material in order to change the firing shrinkage behavior of the conductor and ceramic. For cause (2), it is necessary to revise the laminating conditions in the previous process, or review the organic binder resin that serves as a bond between the materials during lamination. In the case of cause (3), it is necessary to set the firing environment or profile so that the binder spatters easily, however it is necessary to take care that this setting does not adversely affect other materials (ceramic and metal), for example by ensuring that the copper does not oxidize. It is best to avoid returning to the starting material as far as possible, as that necessitates revising all the intermediate process conditions, however the ceramic binder itself may have to be changed in some cases. If (4) is the cause, it is necessary to reconsider not only the firing process but also the combination of materials as well.

This book recognizes that the materials and processes of LTCC technology are interrelated as suggested above. It describes separately the general technical information of each material (ceramic, conductor, and resistor materials) and each process, and it offers commentaries on unique examples resulting from these interrelations. In addition, it includes basic, textbook level content on materials.

Table 1.5 Material specifications of LTCCs of the relevant companies (2000) [Ref. 30].

LTCC Suppliers	Products (Composition)	Di-electric const. (ε)	Qvalue (1/tan δ)	Thermal expansion coefficient (ppm/°C)	Thermal conduct-ivity (w/m • K)	Flexural strength (MPa)
Murata	BAS (Celsian)	6.1	300 (5 GHz)	11.6	2.5	157
	CZG (CaOZrO3+Glass)	25.0	700 (5 GHz)	7.0	2.5	211
	-	60.0	700 (5 GHz)	-	-	-
NEC glass (powder supply)	MLS-25M (Al_2O_3-B_2O_3-SiO_2)	4.7	300 (2.4GHz)	-	-	-
	MLS-1000 (PbO-Al_2O_3-SiO_2)	8.0	500 (2.4GHz)	6.1	-	275
	MLS-41 (Nd_2O_5-TiO_2SiO_2)	19.0	500 (2.4 GHz)	-	-	-
	MLS-61	8.1	150 (2.4 GHz)	7.3	-	255
Sumitomo metal electrodevice	LFC(CaO-Al_2O_3-SiO_2-B_2O_3+Al_2O_3)	7.7	-	5.5	-	270
NEC Vacuum glass	GCS78	7.8	>300 (1 MHz)	-	3.5	250
	GCS71	7.1	>300 (1 MHz)	-	3.2	250
	GCS60	6.0	>300 (1 MHz)	-	1.3	250
NGK	GC-11	7.9	200 (3 GHz)	6.3	3	240
Kyocera	G55	5.7	800 (10 GHz)	5.5	2.5	200
	GL660	9.5	300 (10 GHz)	6.2	1.3	200
Matsushita kotobuki	MKE-100	7.8	500 (1 MHz)	6.1	2.9	245
Niko	NL-Ag II	7.8	>300 (1 MHz)	5.2	3.6	294
	NL-Ag III	7.1	>300 (1 MHz)	5.5	3.5	294
MARUWA	HA-995	9.7		8.1	29.3	400
Dupont	951	7.8	300 (3 GHz)	5.8	3.0	-
	943	7.8	500 (<40GH)	-	-	-
Ferro	A6M	5.9	500 (3 GHz)	7	-	-
Electro-Science Lab	41020-70C	7-8	200 (1 MHz)	7.4	2.5-3.5	-
Heraeus	CT700	7.5-7.9	450 (1 MHz)	6.7	4.3	240
	CT2000	9.1	1000 (450MH)	5.6	-	310

References

[1] "Development of Ubiquitous Service using Wireless Technology", NTT Technical Journal, No. 3 (2003), pp. 6-12.

[2] R. H. Katz, "Adaptation and mobility in wireless information systems," IEEE Personal Comm. 1st Quarter, (1994), pp. 6-17.

[3] K. Oida, "Action for Development of Frequency Resources," The Journal of the institute of Electronics, Information, and Communication Engineers, Vol. 71, No. 5, (1988), pp. 457-463.

[4] "Restructuring System on a Chip Strategy with Package Technology as the New Innovation," NIKKEI MICRODEVICES, No. 189 March (2001), pp. 113-132.

[5] "Activity Around Technology to Embed Devices Internally in PCB's Suddenly Increases," NIKKEI ELECTRONICS, No. 842, March 3 (2003), pp. 57-64.

[6] H. Stetson, "Multilayer Ceramic Technology," Ceramics and Civilization, No. 3, Oct. (1987), pp. 307-322,

[7] W. J. Gyuvk, "Methods of Manufacturing Multilayered Monolithic Ceramic Bodies," U.S. Patent No. 3,192,086, June 1965.

[8] H. Stetson, "Methods of Making Multilayer Circuits," U.S. Patent No.3,189,978, June 1965.

[9] B. Schwartz, "Microelectronics Packaging: II," Am. Ceram. Soc. Bull., Vol. 63, No. 4, (1984) pp. 577-81.

[10] A. J. Blodgett, and D. R. Barbour, "Thermal conduction module: A high performance multilayer ceramic package," IBM J. Res. Develop., Vol. 26, No.3 , May (1982), pp. 30.

[11] C. W. Ho, D.A. Chance, C. H. Bajorek, and R. E. Acosta, "The Thin-Film Module and High Performance Semiconductor Package", IBM J. Res. Develop., Vol. 26, No.3 , May (1982), pp 286-296.

[12] "Low-Temperature Fireable Multi-layer Ceramic Circuit Board", NIKKEI NEW MATERIALS, Aug. 3rd (1987), pp. 93-103.

[13] "High performance and low cost copper paste", NIKKEI ELECTRONICS, Jan. 31st (1983), pp. 97-114.

[14] R. R. Tummala, "Ceramics and Glass-Ceramic Packaging in the 1990s," J. Am. Ceram. Soc., Vol. 74, No. 5 (1991), pp. 895-908.

[15] K. Niwa, E. Horikoshi, and Y. Imanaka, "Recent Progress in Multilayer Ceramic Substrates," Ceramic Transactions Vol. 97, Multilayer Electronic Ceramic Devices (American Ceramic Society, Westerville, OH, 1999)pp. 171-182.

[16] N. Kamehara, Y. Imanaka, and K. Niwa, "Multilayer Ceramic Circuit Board with Copper Conductor", Denshi Tokyo, No. 26 (1987), pp. 143-148.

[17] K. Utsumi, Y. Shimada, T. Ikeda, H. Takamizawa, S. Nagasako, S. Fujii and S. Nanamatsu, "Monolithic Multicomponents Ceramic (MMC) Substrate", NEC Res. & Develop., No. 77, April (1985), pp. 1-12

[18] H. Tsuneno, "Multilayer technology of circuit board" Research and Development of Ceramic Devices and Material for Electronics (CMC, Tokyo, 2000), pp. 128-139.

[19] C. Makihara, M. Terasawa, and H. Wada, "The Possibility of High Frequency Functional Ceramics Substrate" , Ceramic Transactions Vol. 97, Multilayer Electronic Ceramic Devices (American Ceramic Society, Westerville, OH, 1999), pp. 215-226.

[20] P. Chahal, R. R. Tummala, M. G. Allen, and M. Swaminathan, "A Novel Integrated Decoupling Capacitor for MCM-L Technology", Proceeding of 1996 Electronic Components and Technology Conference, (1996), pp. 125-132.

[21] V. Agarwal, P. Chahal, R. R. Tummala, and M. G. Allen, "Improvements and Recent Advances in Nanocomposite Capacitors Using a Colloidal Technique", Proceeding of 1998 Electronic Components and Technology Conference, (1998), pp. 165-170.

[22] S. Ogitani, S. A. Bidstrup-Allen, and P. Kohl, "An Investigation of Fundamental Factors Influencing the Permittivity of Composite for Embeded Capacitor", Proceeding of 1999 Electronic Components and Technology Conference, (1999), pp. 77-81.

[23] H. Windlass, P. M. Raj, S. K. Bhattacharya, and R. R. Tummala, "Processing of Polymer-Ceramic Nanocomposites for System-on-Package Application", Proceeding of 2001 Electronic Components and Technology Conference, (2001), pp. 1201-1206.

[24] S. Nishigaki, S. Yano, S. Fukuta, M. Fukaya, and T. Fuwa, "A New Multilayered Low-Temperature Fireable Ceramic Substrate", 85 International Symposium of Hybrid Microelectronics (ISHM) Proceeding , (1985), pp. 225-234.

[25] Y. Shimada, K. Utsumi, M. Suzuki, H. Takamizawa, M. Nitta, and T. Watari, "Low Firing Temperature Multilayer Glass-Ceramic Substrate", IEEE Transaction on CHMT-6 (4), , April (1983), pp. 382-388.

[26] T. Nishimura, S. Nakatani, S. Yuhaku, T. Ishida, "Co-Fireable Copper Multilayered Ceramic Substrate", IMC 1986 Proceedings, May (1986), pp. 249-271.

[27] H. Mandai, K. Sugoh, K. Tsukamoto, H. Tani, M. Murata, "A Low Temperature Cofired Multilayer Ceramic Substrate Containing Copper Conductors", IMC 1986 Proceedings, May, (1986), pp. 61-64.

[28] K. Niwa, N. Kamehara, K. Yokouchi, and Y. Imanaka, "Multilayer Ceramic Circuit Board with a Copper Conductor", Advanced Ceramic Materials, Vol. 2, No. 4, Oct. (1987) pp. 832-835.

[29] S. Tosaka, S. Hirooka, N. Nishimura, K. Hoshi, and N. Yamaoka, "Properties of a low temperature fired multilayer ceramic substrate", ISHM Proc. (1984), pp. 358.

[30] Advanced LTCC Technology 2001 (Navian, Nagoya, 2001).

Part 1

Material technology

Chapter 2

Ceramic material

2.1 Introduction

LTCCs evolved from HTCCs with the purpose of achieving low loss, high speed, and high density packaging, as higher material performance was required for ceramic material than that offered by the alumina used for HTCCs.

The major characteristic of LTCCs is that metals with low conductor resistance – Cu, Au, Ag and their alloys – are introduced into the ceramic as wiring, thus controlling conductor loss to a low level. As Table 2-1 shows, all the metals with low electrical resistance have a low melting point of around 1,000°C, and in order to allow cofiring with these metals, LTCC ceramics are required to be able to be fired at less than 1,000°C [1, 2].

Table 2-1   Electrical resistance and melting point of conductor metals.

Metal	Electrical resistance ($\mu\Omega \cdot cm$)	Melting point (°C)
Cu	1.7	1,083
Au	2.3	1,063
Ag	1.6	960
Pd	10.3	1,552
Pt	10.6	1,769
Ni	6.9	1,455
W	5.5	3,410
Mo	5.8	2,610

Loss in high frequency circuits (the inverse number of value Q) is expressed as the relationship between dielectric loss and conductor loss ($1/Q_{total} = 1/Q_c + 1/Q_d$, Q_c: conductor Q value, Q_d: dielectric Q value), and the higher the frequency becomes, the greater the effect of dielectric loss over conductor loss [3, 4]. For this reason, ceramics are required to have low dielectric loss.

In high frequency electronic components, several kinds of ceramic with different dielectric constants suited to the function of the circuit are desirable, embedded in a monolithic structure [5, 6]. For transmission lines, a low dielectric constant is effective for achieving high speed transmission of signals (as the propagation delay time of the signal T_{pd} is proportional to the square root of the dielectric constant). On the other hand, the wavelength λ_d of the electromagnetic waves in the dielectric is inversely proportional to the square root of the dielectric constant, so for making compact components such as filters and so on, a high dielectric constant is beneficial. Furthermore, it is necessary to introduce a high dielectric constant layer when forming decoupling functions.

In addition to these characteristics, and in order for components to maintain stable characteristics in their environment of use and for mounted components to retain reliable interconnections, it is important for the ceramics to have low thermal expansion (in particular, they should have a thermal expansion coefficient close to that of the silicon material of the mounted components). Additionally, they must have sufficient strength to withstand the stresses of product assembly during manufacture, as well as while in use. Furthermore, to efficiently release the heat generated by the LSI components mounted on it, ceramic material with high thermal conductivity is desirable.

In order to meet these requirements, composites of glass and ceramic, crystallized glass, composites of crystallized glass and ceramic, and liquid phase sintered ceramics are being developed, as shown in Table 1-1. As examples of liquid phase sintered ceramics, the compounds $BaSnB_2O_6$, $BaZrB_2O_6$, $Ba(Cu_{1/2}W_{1/2})O3$, Bi_2O_3-CuO type, $Pb(Cu_{1/2}W_{1/2})O3$, Bi_2O_3-Fe_2O_3 type, PbO-Sb_2O_3 type, PbO-V_2O_5 type and $Pb_5Ge_3O_{11}$ (melting point: 738°C), LiF, B_2O_3, Bi_2O_3, $Pb_5Ge_{2.4}Si_{0.6}O_{11}$ (melting point: 750°C), Pb_2SiO_4 (melting point: 750°C), $Li_2Bi_2O_5$ (melting point: 700°C) and so on are known as sintering additives that are added to ceramics at less than 10 wt% [7, 8, 9], but they are not typical of LTCCs.

This chapter focuses on the composites of glass and ceramic (especially alumina) that are most commonly used for LTCCs, and it describes the important points for development of materials that meet the requirements for the qualities of LTCCs – firing temperature, dielectric constant, dielectric loss, thermal expansion, strength, and thermal conductivity. This information can also be applied to high dielectric constant LTCCs in which alumina is replaced with perovskite oxides.

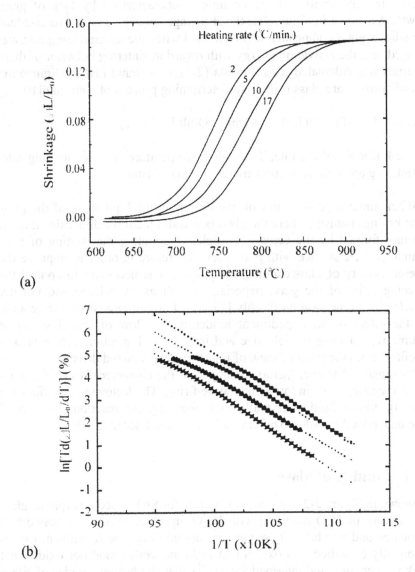

(a)

(b)

Figure 2-1   Curves for linear shrinkage rate (a) and curves for activation energy (b) in alumina/lead borosilicate glass composites at different programming rates [Ref. 10, 11].

2.2   Low temperature firing

The selection of glass materials is very important for sintering glass/ceramic composites, since the liquidation of the glass takes a dominant role in the viscous flow mechanism among the constituents. The ceramic particles in the composites dissolve slightly in the glass during sintering, although the amount is very small and the ceramic is characterized by lack of grain growth. As shown in Figure 2-1, the shrinkage behavior in alumina and lead borosilicate glass composites was examined when the programming rate was changed, and the activation energy with regard to sintering behavior of these materials was calculated with formula (2-1). It is found that the liquidation of lead borosilicate glass is the rate-determining process of sintering [10, 11].

$$\ln[Td(\Delta L/L_0)/dT] = \ln(1/nK_0^{1/n})-1/n\ln\alpha-Q/nRT \qquad (2\text{-}1)$$

$\Delta L/L_0$: linear shrinkage rate, T: absolute temperature, α: programming rate, Q: sintering apparent activation energy, R: gas constant

When sintering composites of glass/ceramic, the liquidation of the glass is the key mechanism, where the glass penetrates the three dimensional mesh structure formed by the ceramic particles, facilitating the wetting of each ceramic particle surface with glass melt. Therefore in order to improve the sintered density of glass/ceramic composites, it is necessary to control the softening point of the glass material, as well as its volume and powder particle size to increase its fluidity [12, 13]. Furthermore, since the ceramic has the effect of an impediment hindering the flow of the glass, using ceramic with a large particle size and thus a small specific surface area is beneficial from the point of view of improving the sintered density.

As suggested above, factors arising from the characteristics of the glass play a very large role in low temperature firing. The following describes the basics of glass – fluidity, crystallization, foaming, and reactions – that need to be understood in order to achieve a high sintered density.

2.2.1 Fluidity of glass

As shown in Figure 2-2, a common structure for SiO_2 based amorphous glass is a network of Si-O modified with Na_2O, in which part of the network is segmented and non-bridging oxygen is formed [14]. The constituent oxides are broadly classified into oxides that make networks, modifier oxides that break the network, and intermediate oxides that can become oxides of either type. Since modifier oxides break the network, they lower the softening point of the glass and increase its fluidity. Table 2-2 details the effect on glass characteristics of the representative types of oxide that are ingredients of glass. Table 2-3 and Figure 2-3 show the composition of Corning's commercial glass and the temperature dependency of glass viscosity [15]. The softening point is the temperature at which viscosity is $10^{7.65}$ poise, and this is used as an index of glass fluidity. Using the above information, it is necessary to control the glass composition, and to choose glass with appropriate fluidity that has the various characteristics required. In general, for glass/ceramic composite type LTCCs, borosilicate glass with a softening point of around 800°C is used.

Table 2-2   The effect of various glass ingredients on glass quality.

SiO_2	A substance that forms the network structures of glass. It has a high melting point and high viscosity. If the silica content in glass is high, the glass has a high transition temperature, low thermal expansion, and excellent chemical durability.
B_2O_3	A substance that forms network structures. Added to the network structure of quartz glass, it reduces viscosity without any negative impact on thermal expansion and chemical durability. It is one of the ingredients of heat-resisting glass and chemical glassware.
PbO	Although it does not form network structures, it can connect SiO_4 tetrahedrons. It is used for glass with a large dielectric constant, refractive index and specific resistance. As it is easily deoxidized, heat treatment in an atmosphere containing oxygen is necessary.
Na_2O	A modifier oxide. It lowers the softening point markedly. Furthermore, it increases the thermal expansion coefficient and ionic conductance. It also reduces chemical durability.
K_2O	A modifier oxide. Although it has the same effect as Na_2O, its K ion is comparatively large and therefore immobile.
Li_2O	A modifier oxide. Although it has the same effect as Na_2O, its Li ion is comparatively small and therefore very mobile. Furthermore, it crystallizes readily.
CaO	A modifier oxide. It prevents the migration of the alkali ion, and therefore the specific resistance and chemical durability of the alkali glass increases. Furthermore, the temperature range for thermal processing is narrowed down.
MgO/ ZnO	A modifier oxide. It has the same effect as CaO (its ionic radius is different).
BaO	This is used instead of PbO. It is cheaper than PbO, with a low degree of hazard.
Al_2O_3	An intermediate oxide. It differs in size from SiO_4 tetrahedrons, but in AlO_4 tetrahedrons it can connect to the network structure. It has the effect of controlling crystallization. Furthermore, since it increases viscosity, it makes melting difficult.

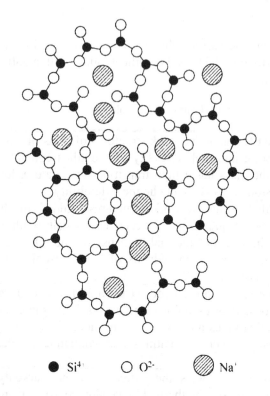

● Si⁴⁺   ○ O²⁻   ◐ Na⁺

Figure 2-2   Structural schema of soda-silicate glass.

2.2.2 Crystallization of glass

In ceramic materials for LTCCs, there are cases where crystals are actively precipitated in the glass to achieve the required characteristics, and cases where precipitation of crystals that are not required in the glass is hindered. In either case, it is necessary to fully understand the occurrence of crystallization of glass, and to control crystal precipitation.

Table 2-3   Composition of commercial glass (wt%).

Glass coating (Corning)	SiO₂	B₂O₃	Al₂O₃	Na₂O	K₂O	MgO	CaO	PbO
FQ (Fused Silica)	99.8	-	-	-	-	-	-	-
Vycor 7900	96	3	1	-	-	-	-	-
Pyrex 7740	81	13	2	4	-	-	-	-
0080	72.6	0.8	1.7	15.2		3.6	4.6	-
0010	63	-	1	8	6	-	1	21
8870	35	-	-	-	7	-	-	58

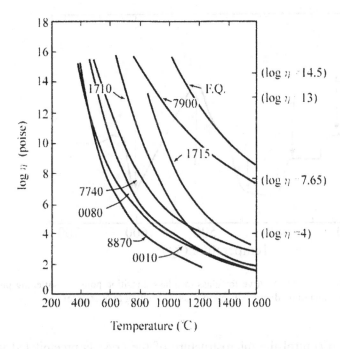

Figure 2-3   Temperature dependency of viscosity of commercial glass and various characteristic temperatures (Strain point: $10^{14.5}$ poise, Annealing point: 10^{13} poise, Softening point: $10^{7.65}$ poise, Working point: 10^4 poise).

The formation of crystals in glass is broadly classified into two types, homogenous nucleation, and heterogeneous nucleation. Homogenous nucleation is where crystal nuclei are formed from a uniform glass phase to precipitate crystals. On the other hand, with heterogeneous nucleation, crystals are precipitated and grow from nuclei formed around elements for forming crystal nuclei introduced into the glass, as well as on the surface of foreign matter present in the glass, at the contact surfaces between the glass and its container such as a crucible, at the surface of the glass and so on. Crystallized glass with TiO_2, ZrO_2, metal ions and the like introduced into the base glass as the nucleation agent is representative of heterogeneous nucleation. With heterogeneous nucleation, differential thermal analysis (DTA) is generally used for analysis of the glass crystallization process and for examination of the crystal precipitation conditions. Figure 2-4 shows a typical DTA curve for glass [16]. The various kinds of characteristic temperature of glass can be estimated from the DTA results as shown in the Figure 2-4.

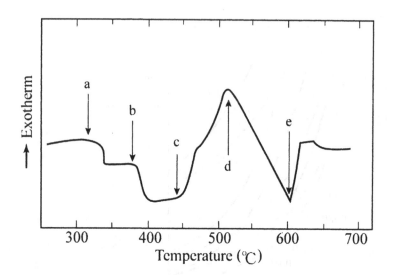

Figure 2-4   A typical DTA curve for glass (a: glass transition point, b: softening point, c: crystallization temperature, d: crystallization peak, e: melting temperature).

In order to control the microstructure of the crystals precipitated in the glass, it is necessary to understand nucleation temperature and nucleation speed, and the temperature for crystal growth and its speed. As in Figure 2-5, nucleation speed I and crystal growth speed U show a Gaussian distribution in relation to temperature, and at a specific temperature they reach their greatest speed. In the case of (a) in the figure, there is a big difference between the nucleation peak temperature and the crystal growth peak temperature, and as neither speed is high, the crystal nuclei formed in the glass disappear before the crystal growth temperature is reached so that vitrification is achieved easily. On the other hand, in the case of (b), I and U are close, and at the temperature range where many nuclei are generated, crystal growth speed is high so that crystals are formed readily in the glass matrix. In order to precipitate many crystals, a method is being tested whereby, after first performing heat treatment at the temperature where nucleation speed is high to generate many crystal nuclei, heat treatment is carried out with a heating schedule that maintains a high crystal growth speed T_u.

The nucleation temperature is decided by obtaining the exothermal peak temperature of crystallization measured by carrying out DTA analysis on each glass that is heat treated at various temperatures in advance (Figure 2-6 (a)), and by calculating the nucleation speed from the temperature difference with the crystallization temperature of glass that is not heat treated. Below are the details of the calculation method for defining the relationship between the shift in the exothermal peak of crystallization when the pre-heating process is carried out, and nucleation speed.

The dynamics of the process in which nucleation and crystal growth occur together are expressed by the Johnson-Mehl-Avrami (JMA) equation.

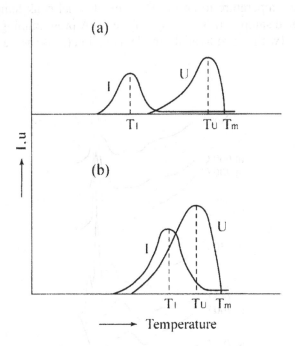

Figure 2-5   The temperature dependency of nucleation speed I and crystal growth speed U [(a): Vitrification occurs readily, (b) Crystals are formed readily] [Ref. 17].

-ln (1-x) = $(kt)^n$ . . . . (2-2)
x: Volume fraction of crystals

k = Aexp (-E/RT) . . . . (2-3)

A: Constant, E: Activation energy of crystal growth, $N = N_r + N_0/\alpha$, N_r is the number of crystal nuclei at the unit volume formed during heat treatment at nucleation temperature, and N_0/α is the number of crystal nuclei at the unit volume formed during the rise in temperature at speed α.

If the temperature rises at the constant programming rate α, since k changes in accordance with the temperature or time, equation (2-2) is expressed as equation (2-4).

-ln $(1-x) = \{1/\alpha \int k\,(T)\,dT\}^{n}$ .... (2-4)

If equation (2-3) is substituted with (2-4), and the integral is found, and additionally the logarithm is taken, the following equation is obtained.

$(1/n) \ln \{-\ln(1-x)\} = \ln (N_r+N_0/\alpha) - \ln\alpha - 1.052E/RT + \text{const.}$ .... (2-5)

Assuming the temperature to be the DTA exothermal peak temperature T_p, for a non-treated sample, if $N_r = 0$, and if the DTA programming rate is fast, since $N_r/ N_0 \gg 1$ with a heat treated sample, equation (2-5) can be rewritten as (2-6).

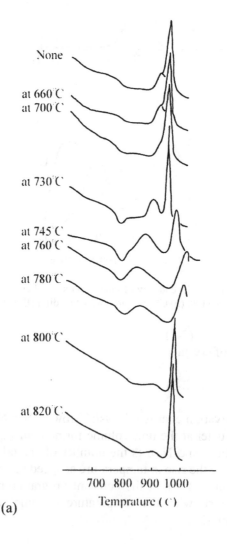

(a)

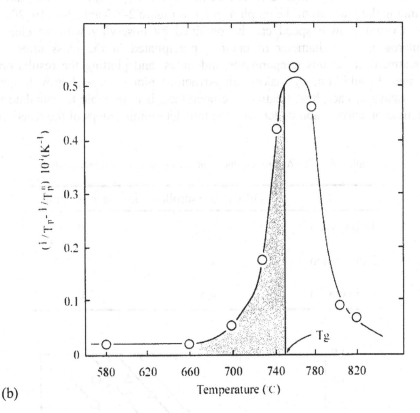

(b)

Figure 2-6   DTA curves for non heat treated glass and for glass that is heat treated at various temperatures for 8 hours (a), and the relationship between heat treatment temperature and nucleation speed (b) [Ref. 18].

$$\ln N_r = 1.05(E/R)[1/T_p - 1/T^0_p] + \text{const.} \quad \ldots \ (2-6)$$

where, T^0_p is the exothermal peak temperature of the non-treated sample, while T_p is the DTA exothermal peak temperature of the sample after heat treatment at the nucleation temperature. If nucleation processing is carried out under isothermal conditions, then $N_r = It^b$. Here, I is nucleation speed, b is the constant, and t is the isothermal processing temperature. If the heat treatment time with a constant time of $N_r = It^b$ is replaced with (2-6), $\ln I = 1.052(E/R)[1/T_p - 1/T^0_p] + \text{const.}$

In this way, the heat treatment temperature dependency of nucleation speed can be expressed with variations of $[1/T_p - 1/T^0_p]$.

Furthermore, the crystal precipitation mechanism can be established by obtaining the amount of crystallization of the glass after testing with heat treatment at various temperatures and times, plotting a graph in accordance with the JMA equation rewritten as $\ln\{\ln[1/(1-x)]\} = n\ln k + n\ln t$, and by finding the gradient of the graph n (refer to Figure 2-7, Table 2-4)[19, 20].

Crystal growth speed can be obtained by observing with an electron microscope the diameter of crystals precipitated in the glass after heat treatment at various temperatures and times, and plotting the results on a graph. In addition, by making an Arrhenius plot of crystal growth speed constants at each heat treatment temperature, it is possible to calculate the activation energy, and understand the rate determining steps of the reaction.

Table 2-4   The Avrami exponent for each form of crystal precipitation.

	Diffusion controlled	Interface controlled
3-dimension	1.5	3.0
2-dimension	1.0	2.0
1-dimension	0.5	1.0

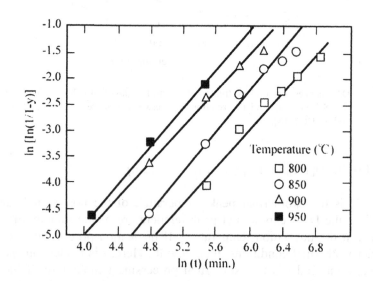

Figure 2-7   Plot of $\ln\{\ln[1/(1-x)]\}$ vs. nlnt for crystals precipitated in heat treated glass at 800°C to 950°C [Ref. 19, 20].

2.2.3 Foaming of glass

The formation of internal pores that is observed in LTCCs is sometimes caused by insufficient sintering, and sometimes by excessive sintering causing the occurrence of gas within the material. Figure 2-8 shows the microstructure in glass/alumina composite material. The pores from insufficient sintering observed in a sample fired at 800°C are angular (a), while the pore shape assumes a roundness (b) with increased firing temperature. Pores due to excessive sintering in material fired at 1,100°C appear spherical (c). There are two possible causes of these spherical pores. Cause (1): During sintering, the surface of the sample is sintered first, and after a well-sintered film is formed on the surface, the pores are formed when gas left inside the material or residues of organic binder are expelled at high temperature. Cause (2): The gas dissolved in the raw glass powder used in LTCCs is released at high temperature forming the pores.

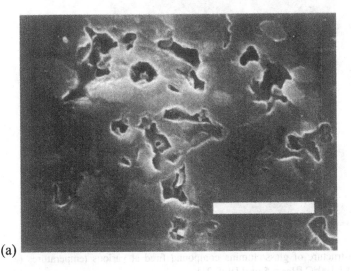

(a)

When glass melts, decomposition of the batch materials such as H_2BO_3, Na_2CO_3, Na_2SO_4, $NaNO_3$ and so on, releases large volumes of gas such as CO_2, SO_2 and the like[22, 23, 24, 25]. Most of this is released, however some of the gas forms bubbles and remains in the glass or dissolves inside the glass melt. In order to prevent foaming, it is important to examine the glass raw powder and to use raw materials containing little dissolved gas. (Caution is required as some commercially available glass powder contains ground up reject glass products.) In addition, reducing the time during the firing process at temperature ranges where gas occurs readily is effective in controlling the foaming of glass.

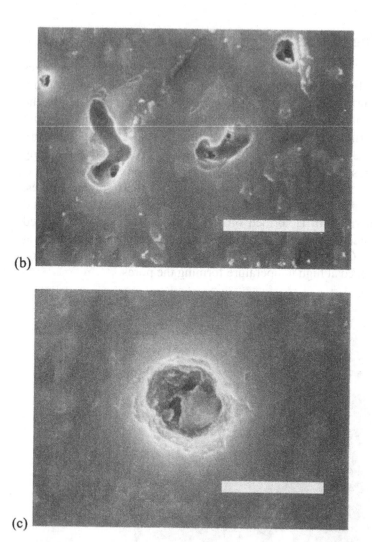

(b)

(c)

Figure 2-8   Microstructure of glass/alumina compound fired at various temperatures (a) 800°C, (b) 900°C, (c) 1,100°C [Bar = 5 μm] [Ref. 21].

Figure 2-9 shows the results of an investigation of changes in sintered density of glass/alumina composite at a lower temperature (900°C) than the final firing temperature of 1,000°C and when the retention time was changed. Sintered density falls abruptly with retention time. Although pores are not observed in the surface of the ceramic, many of the spherical pores noted above can be seen inside the ceramic like foam glass (Refer to Figure 2-10).

These results substantiate the view that gas originating inside the material forms the spherical cavities.

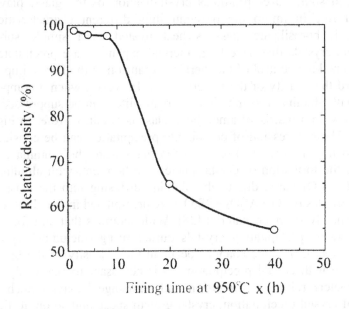

Figure 2-9   Sintered density of glass/alumina composite when 950°C retention time is changed.

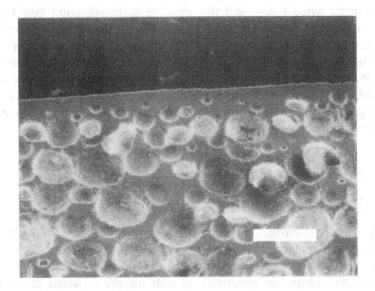

Figure 2-10   Microstructure of glass/alumina composite after heat treatment of 950°C for 20 h, and 1,000°C, for 5 h [Bar = 200 μm].

2.2.4   Reaction between glass and ceramic

In glass/alumina composites, the amount of alumina dissolved in the glass at firing is small, however this small amount suppresses crystallization of the glass, or in some cases promotes crystallization of the glass, playing an important role in improving or controlling different characteristics. For example, if borosilicate glass is heat treated as a simple substance, cristobalite crystals that have large thermal expansion are precipitated, and as well as making control of the thermal expansion of the LTCC impossible, they retard the density of the material [26]. However, when a composite is formed with alumina, precipitation of cristobalites can be suppressed, and a composite with a matrix of amorphous glass is obtained (refer to Figure 2-11) [27]. The suppression of cristobalite precipitation can be considered to be due to the alumina diffused into the glass from the alumina particles hindering the formation of crystal nuclei. Furthermore, with alumina/CaO-Al_2O_3-SiO_2-B_2O_3 glass, due to the alumina diffusing into the glass during firing, anorthites (CaO • Al_2O_3 • $2SiO_2$) are precipitated in the glass resulting in mechanically stronger material [28]. With ceramics that aim for extreme precision while precipitating crystals during firing (including crystallized glass type LTCCs), since changes occur in the viscosity of the base glass phase along with crystal precipitation, it is necessary to rigorously control the parameters related to precision and shrinkage behavior such as the amount of crystal precipitation, crystal growth speed and so on, to fire with good repeatability. Furthermore, if warping occurs when firing the substrate, since the amount of remaining glass phase in the ceramic after firing is different from when firing, and the apparent softening point gets higher, unlike glass/ceramic composite types, the warping cannot easily be fixed by reheating and it will be necessary to control the processing conditions of the firing process more exactly. If LTCCs are used for circuit boards with fine wiring formed internally, since control of the wiring circuit dimensions is important, LTCCs of the amorphous glass and ceramic composite type with their simple sintering density process are beneficial from the point of view of shrinkage dimension control.

2.3   Dielectric characteristics

2.3.1   Dielectric constant

Since LTCCs are basically composite structures of glass and crystals, controlling their dielectric constant depends largely on the combination of constituent materials of the composite and its material composition (volume fraction of the constituent materials). In addition, the dielectric constant of the constituent materials themselves (especially the glass material) has a big influence on the dielectric constant of the LTCC.

The dielectric constant of the materials themselves depends on the contribution of electrons or ions with regard to polarizability and their dipole orientation, and the following relationship obtains between polarizability $N\alpha$ and relative permittivity ε per unit volume.

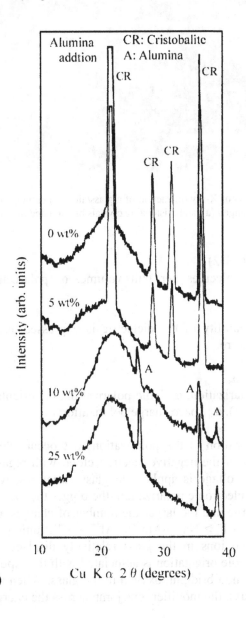

(a)

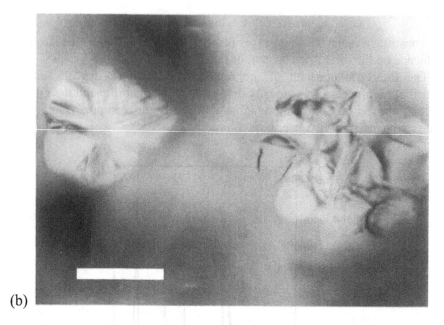

(b)

Figure 2-11   Results of X-ray diffraction of a glass/alumina composite when the amount of alumina added is changed (a), and cristobalite crystals precipitated in glass (b) [Bar = 1 μm].

$$N\alpha/3\varepsilon_0 = (\varepsilon - 1)/(\varepsilon + 2)$$
N: number of molecules per unit volume, α: polarizability, ε: relative permittivity,

The total polarizability of the dielectric is expressed as the sum of each polarizability feature.

$$\alpha = \alpha_e + \alpha_i + \alpha_o + \alpha_s$$
α_e: electronic polarization, α_i: ionic polarization, α_o: orientation polarization (dipole orientation), α_s: space charge polarization

Electronic polarization is the polarization that occurs due to the shift in center of gravity of the negative electron cloud with regard to the positive nucleus when a voltage is applied. In glass structures, with regard to the polarizability of electronic polarization, the bigger the ionic radius, the more negative the charge and the larger the number of charges as in $Ba^{2+} > Sr^{2+} > Ca^{2+} > Mg^+$, $O^{2-} > F^- > Na^+ > Mg^{2+} > Al^{3+} > Si^{4+}$. Ionic polarization occurs when the positive ions in the glass relatively displace the anions in an electric field. Dipole orientation is associated with the dipoles formed of the modifier ion and non-bridging oxygen in the glass. When an electric field is applied to the glass, the modifier ions jump across the energy barrier formed in the vicinity, and migrate in the direction of the electric field. If this change in the electric field is slow, the ions jump across a high energy barrier and migrate over a long distance, however, when the change in the electric field is fast, they jump across a low energy barrier, and do not migrate far. For this reason, it is a mechanism that occurs in low frequency regions. Dipole orientation is large in glass that includes alkali ions and OH - ions. Space charge polarization is polarization of migrated charges that accumulate in the vicinity of an electrode, grain boundaries within the material, and at the interfaces of dissimilar materials, without being neutralized. Seen from an external circuit, it appears as though capacity has increased. There are methods using a space charge, as a technology for achieving a high dielectric constant. As in Figure 2-12 by performing reduction processing and valency control of the basic dielectric particles themselves, they are made into semiconductors and their apparent dielectric constant is increased, and by giving the grain boundary part insulating properties, the withstand voltage is improved [29]. Another method involves, conversely, forming a material with high conductivity in a grain boundary phase, in a structure made up of dielectric particles with excellent insulation [30].

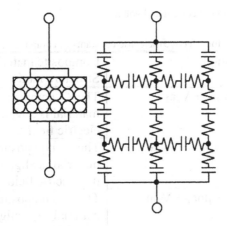

Figure 2-12   Microstructure model of a barrier layer type dielectric and its equivalent circuit.

In the composite, the dielectric constant is determined by the dielectric constant and volume fraction of the constituent material, and the complex form of the constituent material [31]. Table 2-5 shows the 4 different models of mixing rules for complex forms of the constituent materials. LTCC ceramics, being of the type with ceramic particles distributed in a glass matrix, fit the Maxwell model well. In order to achieve a lower dielectric constant material, approaches are being made such as introducing relative permittivity 1 air phase as constituent material, for example with the use of hollow ceramic as a constituent material, forming a cavity phase during the firing process.

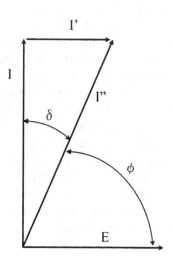

Figure 2-13   Phase relation of current and voltage.

Table 2-5   Mixture rules for effective dielectric constants of composites.

Composite type	Equation	Composite material Structure model
Parallel	$E = V_1\varepsilon_1 + V_2\varepsilon_2$	The composite constituent material is aligned parallel to the electric field
Series	$1/\varepsilon = V_1/\varepsilon_1 + V_2/\varepsilon_2$	The composite constituent material is aligned in series with the electric field
Logarithmic	$\ln\varepsilon = V_1\ln\varepsilon_1 + V_2\ln\varepsilon_2$	The composite constituent material is aligned randomly (empirical rule)
Maxwell	$\varepsilon = \{V_2\varepsilon_2\,(2/3 + (\varepsilon_1/3\varepsilon_2)) + V_1\varepsilon_1\}/\{V_2(2/3 + (\varepsilon_1/3\varepsilon_2)) + V_1\}$	Spherical phases are distributed in the matrix (0-3 connectivity)

ε: dielectric constant of the composite, ε_1: dielectric constant of constituent material 1, ε_2: dielectric constant of constituent material 2, V_1: volume fraction of constituent material 1, V_2: volume fraction of constituent material 2

2.3.2 Dielectric loss

When an alternating voltage is applied to a capacitor that does not include a dielectric, power loss does not occur since the phase of the current is 90° in front of the voltage. However, if an electric field is applied to a capacitor that includes a dielectric, there is a phase shift in the electric displacement of the electric field, and some of the electric energy changes into heat in the dielectric. Dielectric loss is the amount of electric energy lost through conversion to heat in the dielectric when an electric field is applied. In Figure 2-13, the phase angle Φ of current I' is 90° smaller, and accompanies voltage E and the in-phase current component I''. Therefore, loss corresponding to I'' occurs. The size of this loss is expressed as tan δ = I''/I using δ as the complement of Φ.

In order to reduce the dielectric loss of LTCCs, similarly with the dielectric constant, it is effective to construct them of materials with low dielectric loss, use a large volume fraction of low dielectric loss materials, and to make the dielectric loss of each constituent material small. Among glass, a constituent material of LTCCs with large dielectric loss, the following four dielectric loss mechanisms are known. (1) Conduction loss through electric conductivity, (2) dipole relaxation loss from relaxation necessitated when the alkali ion, OH - ion and so on reciprocate between the adjacent position due to the electric field, (3) distortion loss when the network structure of the glass distorts due to the electric field, and dipole orientation occurs momentarily, and (4) ion vibration loss caused when there is resonance at the proper oscillation frequency decided by the mass of structural ions and the chemical bonding strength of the surroundings. If alkali is substituted with ions that have a large ionic radius such as barium and the like in lossy glass that includes ions with high ionic mobility such as alkali ions, loss can be reduced since the mobility of the ions can be hindered [32, 33, 34].

While dielectric characteristics in the microwave band are determined by ionic polarization and electronic polarization, dielectric loss through electronic polarization is small enough to be ignored, and the following equation can be derived from the one-dimensional lattice vibration model through ionic polarization (qualitatively extensible to three-dimensional ion crystals).

$$\tan \delta = (\gamma/\omega_T^2) \, \omega$$

ω_T : resonating angular frequency of the optical mode of lattice vibration transverse waves, γ: attenuation constant, ω: angular frequency

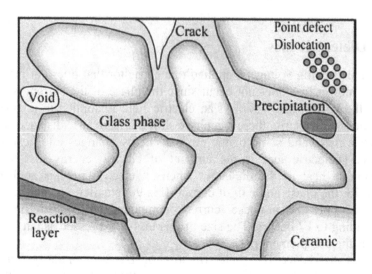

Figure 2-14   Micro and macro defects inherent in LTCCs.

As the presence of lattice defects, impurities and grain boundaries are factors that increase γ, it is effective to use raw materials with high purity to achieve low dielectric loss, and to aim for a microstructure without impurities and without the internal micro and macro flaws that are shown in Figure 2-14 [35].

2.4   Thermal expansion

Several kinds of mixing rules are presented in Table 2-6 regarding the thermal expansion coefficient of composites. By predicting the thermal expansion coefficient using these equations, and by controlling material composition and formulation, it is possible to approach the desired values.

The simple model is a formula for calculating just the mixing ratio, while the Turner model takes into account the effect of isotropic stress of the adjacent phase. The Kerner model is a model that allows for the shear effect of the phase boundary in types with spherical phases distributed isotropically in the matrix, and it shows values between those of the Turner model and the simple model. Table 2-7 shows the calculated values using the Turner model [36, 37], and the actual measurements for glass/ceramic composites using different ceramics. In types where a difference can be seen between the value predicted by the calculation and the actual measurement, precipitation of a secondary crystal phase (cristobalite) is identified. The cristobalites formed when the glass crystallizes in a composite, arise from a phase transition between 100 to 200°C, and thermal expansion changes markedly (refer to Figure 2-15)[38, 39]. Heating at around 200°C is required in the assembly process of the substrates, and extreme changes in thermal expansion cause connection failures in the interconnects of mounted components that harm the reliability of the product. Figure 2-16 shows the phase formation of the glass phase in a glass/alumina composite when the amount of alumina and the firing temperature are changed. When a little alumina is added and the firing temperature is low, cristobalite crystals are precipitated in the glass phase, and when a lot of alumina is added and the firing temperature is high, mullite crystals are precipitated. In the wide composition and firing temperature range that remains, the glass phase is amorphous, the glass/alumina composite is stable, and cristobalite crystals are suppressed, so it is demonstrated that the thermal expansion coefficient of glass/alumina composites can be controlled. Besides the additives shown in Table 2-7, spinel ($Al_2O_3 \cdot MgO$) containing the ingredient Al_2O_3 has also been identified as an additive that suppresses cristobalite crystals [40].

Table 2-6   Mixture rules for thermal expansion coefficients of composites.

Simple mixture rule	$\alpha = \alpha_1 V_1 + \alpha_2 V_2$
Turner equation	$\alpha = (\alpha_1 V_1 K_1 + \alpha_2 V_2 K_2)/(V_1 K_1 + V_2 K_2)$
Kerner equation	$\alpha = \alpha_1 + V_2(\alpha_1 - \alpha_2) \cdot \{K_1 (3K_2 + 4G_1)^2 + (K_2 - K_1)(16G_1^2 + 12G_1 K_2)/(4G_1 + 3K_2)[4V_2 G_1(K_2 - K_1) + 3K_2 K_1 + 4G_1 K_1]$

α_1: thermal expansion coefficient of constituent material 1, α_2: thermal expansion coefficient of constituent material 2, V_1: volume fraction of constituent material 1, V_2: volume fraction of constituent material 2, K_1: volume modulus of constituent material 1, K_2: volume modulus of constituent material 2, G: shear modulus

Table 2-7   Thermal expansion coefficients ($\times 10^{-6}/°C$) of each type of glass/ceramic composite (Calculated value using Turner equation and the actual measurement).

Material type	Calculated value	Measured value
Alumina (Al_2O_3) - glass	4.6	4.0
Aluminum nitride (AlN) - glass	3.6	3.5
Mullite ($3Al_2O_3 \cdot 2SiO_2$) - glass	4.8	3.9
Forsterite ($2MgO \cdot SiO_2$) - glass	5.7	18.2
Steatite ($MgO \cdot SiO_2$) - glass	4.6	17.4
Magnesium oxide (MgO) - glass	5.7	18.3
Silicon nitride (Si_3N_4) - glass	3.1	17.1

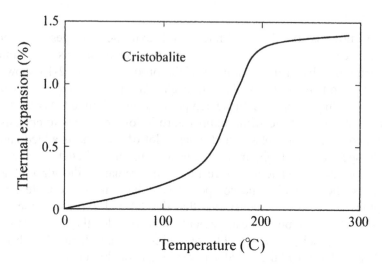

Figure 2-15   Cristobalite thermal expansion curve.

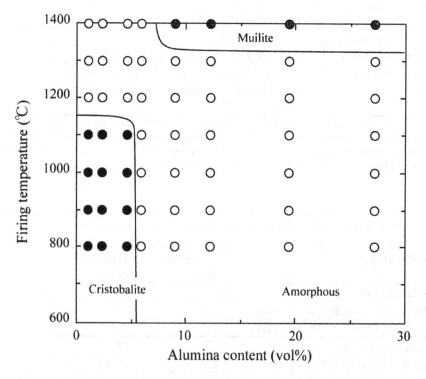

Figure 2-16   Phase formation when the composition and firing temperature of glass/alumina composite are taken as the parameters.

2.5 Mechanical strength

In glass/ceramic composites with distributed ceramic particles, mechanical strength varies according to (1) composition (amount of ceramic), (2) porosity, and (3) ceramic particle diameter. Figure 2-17 shows the flexural strength of a glass/alumina composite when each parameter is changed. In cases where the composition (amount of ceramic) is the parameter, in accordance with the simple mixing rule, strength is seen to improve along with the increase in the amount of alumina [41]. The relationship between porosity and strength nearly fits the following equation proposed by Ryskewitsch, and strength can be predicted using this equation; $\sigma = \sigma_0 \cdot \exp(-np)$, (n: constant, p: porosity) [42]. The flexural strength of glass/ceramic composites when the dispersed particle diameter (d) is varied closely matches the Orowan and Hall-Petch relationship ($\sigma \propto d^{1/2}$) [43, 44], and stronger ceramic can be achieved with the use of alumina with a fine particle diameter. The strength achieved by using finer ceramic particles in glass/ceramic composites is due to their ability to spend the energy required to transmit cracks by dissipating cracks that occur and making them zigzag (boarding and deflection of cracks) (refer to Figure 2-18). With the following management of parameters, it is possible to control the strength of the glass/ceramic composite itself.

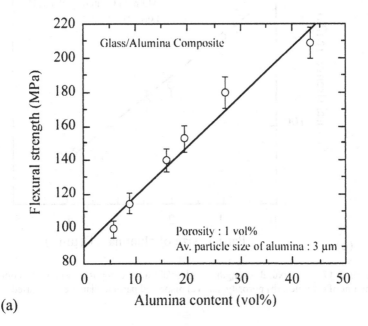

(a)

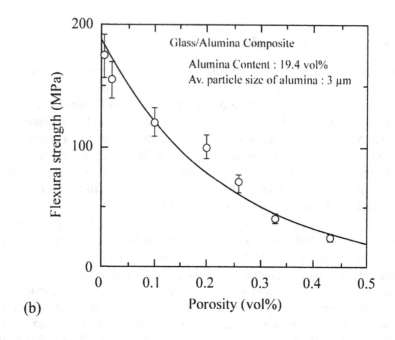

(b)

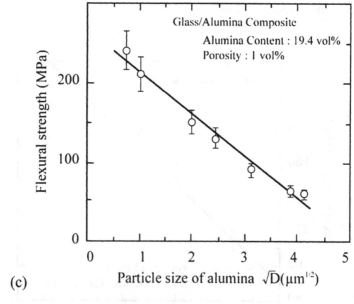

(c)

Figure 2-17   Flexural strength of glass/alumina composite when (a) composition (amount of ceramic), (b) porosity, and (c) dispersed particle diameter are varied.

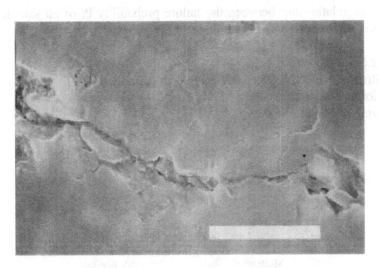

Figure 2-18 Transmission of cracks in glass/alumina composite [Bar = 10 μm].

2.5.1 Strengthening the glass phase

Strengthening a weak glass phase is effective in order to strengthen glass/ceramic composites. The following two methods are known to strengthen the glass phase. (1) Crystallized glass method, and (2) ion exchange strengthening method

With the crystallized glass method, crystals with low thermal expansion formed due to compressive stress are precipitated in the amorphous glass matrix. There are two methods, one using glass material that crystallizes readily, and the other by precipitating crystals by promoting a reaction between the alumina and glass during the firing process.

In the ion exchange strengthening method as shown in Table 2-8, the glass is immersed in molten salt containing potassium ions, and by introducing larger potassium ions into the surface of the glass in place of the sodium ions, the glass network expands, resulting in increased strength due to the compressive stress created (refer to Figure 2-19 for the strength principle) [45, 46]. When borosilicate glass/alumina composite is immersed for 30 hours in KNO_3 at 400°C, as shown in Figure 2-20, it is found that potassium penetrates the surface to a depth of around 100 μm, and bending strength is improved by more than 50% (non-treated: 150 MPa, after ion exchange: 230 MPa).

Variation in the strength of ceramic occurs readily and to evaluate this variation, the Weibull modulus is frequently applied. The Weibull modulus is found as follows.

The relationship between the failure probability P_s of all samples and stress σ when they break is as follows [47].

$$P_s = \exp[- V((\sigma - \sigma_u)/\sigma_0)^m]$$
σ_u: Stress at which breakage does not occur (usually 0)
σ_0: Normalization constant
m: Weibull modulus

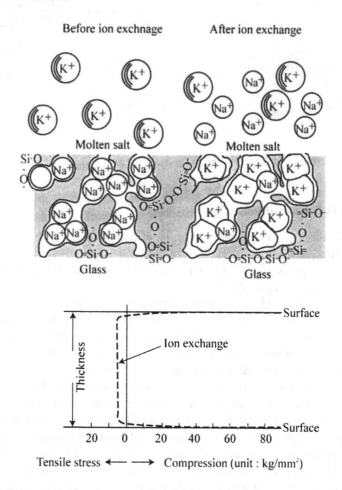

Figure 2-19 Principle of strengthening of the glass due to $Na^+ \rightarrow K^+$ exchange.

If we take the double logarithm of the above equation,
$$\ln\ln(1/P_s) = \ln V + m\ln(\sigma - \sigma_u) - m\ln \sigma_0$$
By plotting the equation on a Weibull modulus sheet and finding the gradient of the straight line, the Weibull modulus can be found. The

measured strength values around N for all samples are arranged from the highest value, and breaking at i is defined as $P_s = 1-i/(N+1)$. Materials with a big Weibull modulus (a big gradient on the Weibull plot) have little variation in strength.

Table 2-8   Physical properties of various kinds of potassium salt.

Potassium salt	K content (%)	Melting point (°C)	Water solution
KNO_3	83	330	Neutral
KCl	52	776	Neutral
K_2SO_4	65	1, 067	Neutral
K_3PO_4	55	1, 340	Alkaline
K_2HPO_4	45	807	Alkaline
K_2CO_3	57	891	Alkaline
KOH	70	360	Alkaline

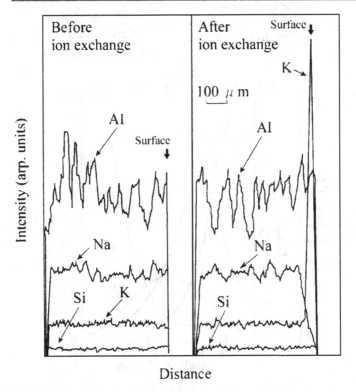

Figure 2-20   EPMA results of a cross section of glass/alumina composite before and after ion exchange treatment (immersion in KNO_3 at 400°C for 30 h).

   With device level LTCCs, the glass/ceramic composites themselves are
not used separately, but are formed with conductor wiring in each layer and
via conductors between layers. Seen from a macro point of view, they can be
characterized as composite materials formed of wiring metal and ceramic.
For this reason, in order to improve the strength of the LTCC overall, it is
effective to reduce micro and macro flaws in the metal/ceramic interface,
and to predict and improve strength, with reference to the equation suggested
with fiber reinforced resin material below [48].

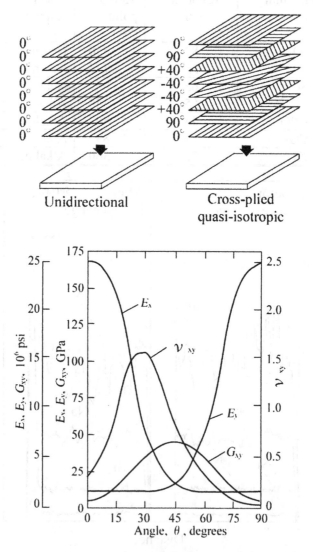

Figure 2-21 The various changes in mechanical properties when the arrangement angle of
carbon fibers in an epoxy/carbon fiber composite laminate is varied.

$E_c = E_f V_f + E_m(1 - V_f)$ (for constraint strain)
$1/E_c = V_f/E_f + (1 - V_f)/E_m$ (for constraint stress)
E_c: modulus of composite, E_f: modulus of fiber, E_m: modulus of matrix,
V_f: fiber volume fraction

Figure 2-21 shows an example where the arrangement angle of carbon fibers in an epoxy/carbon fiber composite laminate is varied in order to change its mechanical properties when laminated. However, if we consider the use of the LTCC wiring in place of the carbon fibers, the mechanical properties of the whole LTCC can be improved by arranging the wiring at an angle. However, there are problems with screen printing diagonal wiring and further process technology development is required for its application.

2.5.2   Thermal shock resistance

When heating a material, the thermal expansion of the high temperature side is greater than that of the low temperature side, and the compressive stress of the heated surface pulls on the inside of the material on the low temperature side causing stress. Conversely, when cooling, the pull is towards the surface causing stress. In this way, when heating and cooling materials, thermal stress (compressive and tensile stress) occurs. Since the compressive strength of ceramics is markedly greater when compared with their tensile strength, cracks start in the weakest parts of those parts where tensile stress is present. The thermal shock resistance coefficient, which is the index for thermal stress σ and thermal shock resistance, is expressed with the following equation [49, 50].

$\sigma = (E\alpha\Delta T)/(1 - \mu)$
μ: Poisson's ratio, E: elastic modulus, ΔT: temperature difference $(T - T_0)$
α: thermal expansion coefficient

With regard to the thermal shock of abrupt cooling from a high temperature, the temperature difference between the initial temperature and the temperature at which cracks start forming in the material ΔT_{max} is called the thermal shock resistance coefficient, and the higher this value, the stronger the resistance to shock.

$\Delta T_{max} = R = \sigma_f(1 - \mu)/E\alpha$
σ_f: fracture strength,

When heating and cooling is slow, thermal conductivity k is added, and thermal shock resistance is defined as R'.

$$R' = \sigma_f(1 - \mu)k/ (E\alpha)$$

Furthermore, if heating and cooling is performed at a certain speed, and thermal diffusivity ($\delta = k/\rho C$) is inserted, thermal shock resistance R'' is defined with the following equation.

$$R'' = \sigma_f(1 - \mu)(k/\rho C)/ (E\alpha) = R'/(\rho C)$$

Table 2-9   Physical properties of thermal shock resistance of LTCCs and their constituent materials.

	Glass/alumina composite	Alumina	Borosilicate glass (Pyrex)
Young's modulus E ($\times 10^6$ MPa)	0.093	0.38	0.07
Poisson's ratio μ	0.17	0.24	0.16
Thermal expansion coefficient α ($\times 10^{-6}/°C$)	4.1	8.0	3.0
Fracture strength σ_f (MPa)	150	300	80
Thermal shock resistance R	326	75	320
Fracture critical temperature T_{max} (°C)	500	200~300	--
Thermal conductivity k (W/mK)	2.5	16	0.96

Table 2-9 shows the related physical properties of thermal shock resistance of LTCCs and their constituent materials, while Figure 2-22 shows fracture strength of a glass/alumina composite after thermal shock is applied at various temperature differences, and its material microstructure after breaking. As shown in Table 2-9, the thermal shock resistance of glass/alumina composites is greater than that of alumina, and it has excellent reliability with regard to the heat history undergone during the various kinds of assembly.

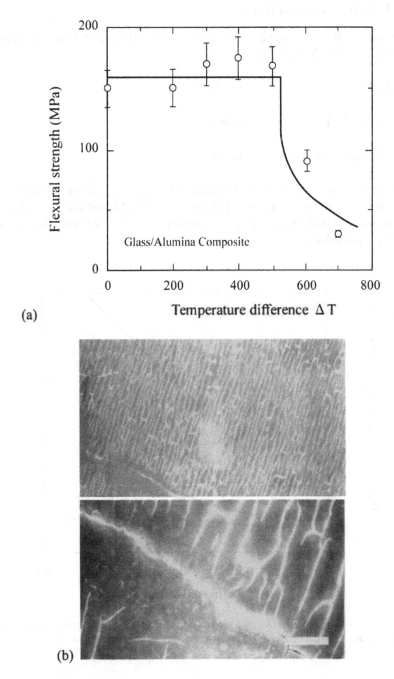

(a)

(b)

Figure 2-22 Thermal shock resistance of glass/alumina composite (a), and material microstructure after thermal shock fracture (b)
[Bar = 20μm].

2.6   Thermal conductivity

The mixture rule for thermal conductivity normally applied for composite materials is shown below [51].

$$k = V_1 k_1 + V_2 k_2 \qquad \ldots \ldots (2\text{-}7)$$

$$1/k = V_1/k_1 + V_2/k_2 \qquad \ldots \ldots (2\text{-}8)$$

$$\log k = V_1 \log k_1 + V_2 \log k_2 \ldots \ldots (2\text{-}9)$$

k: thermal conductivity of the composite material, k_1: thermal conductivity of constituent material 1, k_2: thermal conductivity of constituent material 2, V_1: volume fraction of constituent material 1, V_2 : volume fraction of constituent material 2

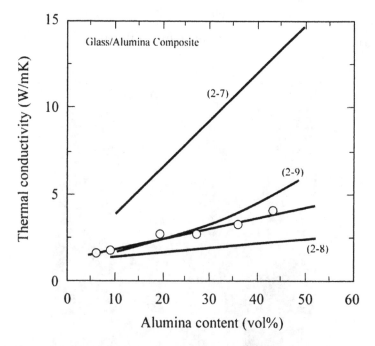

Figure 2-23   Thermal conductivity of glass/alumina composite (actual measurement and calculated value). The numbers in the Figure correspond to the formula numbers.

Equation (2-7) is the mixing rule when heat flow speed direction and the constituent materials are arranged in parallel, and equation (2-8) is when heat flow speed direction and constituent materials are arranged perpendicularly. Glass/ceramic composites of the type in which ceramic particles are distributed in a glass matrix make a good match with the empirical logarithmic law displaying intermediate values of both. Figure 2-23 shows the actual measurement and the values applied to these formulae for a borosilicate glass/alumina composite. When the thermal conductivity of the alumina and borosilicate glass are 28 W/mK and 1.3 W/mK respectively, the thermal conductivity of the glass/alumina composite with alumina content of 19.4 vol% is 2.7 W/mK, and this shows a good approximation to the calculated value of 2.4 W/mk using the logarithmic law.

In order to achieve high thermal conductance, aluminum nitride, silicon nitride, silicon carbide and the like which have high thermal conductance are being tried as ceramic ingredients. However whatever the type, it is difficult to exceed that of the alumina used in HTCCs. However, compared with resin materials, glass/alumina composites can achieve thermal conductivity more than ten times higher.

References

[1] N. Kamehara, Y. Imanaka, and K. Niwa, "Multilayer Ceramic Circuit Board with Copper Conductor", Denshi Tokyo, No. 26, (1987), pp. 143-148.

[2] R. R. Tummala et al., "High performance glass-ceramic/copper multilayer substrate with thin-film redistribution", IBM J. Res. Develop., Vol. 36, No. 5 Sep. (1992), pp. 889-904.

[3] System Design: Say good-bye to PCI, say hello to serial interface, NIKKEI ELECTRONICS, 2001. 6. 18., no. 798 , pp. 119.

[4] Y. Usui, "Quantitative Analysis Overcomes Design Bottleneck for PCB's with Speeds over 1GHz", NIKKEI ELECTRONICS, 2002. 1. 7, pp. 107-113.

[5] A. A. Mohammed, "LTCC for High-Power RF Application?", ADVANCED PACKAGING, Oct. (1999), pp. 46-50.

[6] D. I. Amey, M. T. Dirks, R. R. Draudt, S. J. Horowitz, and C. R. S. Needs, "Opening the door to wireless innovations", ADVANCED PACKAGING, Mar. (2000) , pp. 37-540.

[7] T. Hayashi, T. Inoue, and Y. Akiyama, "Low-Temperature Sintering and Properties of (Pb, Ba, Sr)(Zr, Ti, Sb)O_3 Piezoelectric Ceramics Using Sintering Aids", Jpn. J. Appl. Phys. Vol. 38, (1999), pp. 5549-5552.

[8] K. Murakami, D. Mabuchi, T. Kurita, Y. Niwa, and S. Kaneko, "Effects of Adding Various Metal Oxides on Low-Temperature Sintered Pb(Zr, Ti)O_3 Ceramics", Jpn. J. Appl. Phys. Vol. 35, (1996), pp. 5188-5191.

[9] W. A. Schulze and J. V. Biggers, "Piezoelectric Properties of $Pb_5Ge_3O_{11}$ Bonded PZT Compositions", Mat. Res. Bull., 14, (1979), pp. 721-30.

[10] J. H. Jean and T. K. Gupta, "Isothermal and Nonisothermal Sintering Kinetics of Glass-Filled Ceramics", J. Mater. Res., Vol. 7, No. 12, (1992), pp. 3342-48.

[11] C. R. Chang and J. H. Jean, "Camber Development during Cofiring an Ag-based Ceramic-Filled Glass Package", Ceramic Transactions Vol. 97, (1999), pp. 227-239.

[12] G. C. Kuczynski and I. Zaplatynskyj, "Sintering of Glass", J. Am. Ceram. Soc. , Vol. 39, No. 10, (1956), pp. 349-350.

[13] I. B. Cutler and R. E. Henrichsen, "Effect of Particle Shape on the Kinetics of Sintering of Glass", J. Am. Ceram. Soc. , Vol. 51, No. 10, (1968) pp. 604-05.

[14] W. J. Zachariasen, J. Am. Ceram. Soc., Vol. 54, (1932) pp. 3841.

[15] Corning , "The Characterization of Glass and Glass-Ceramics"

[16] D. Clinton, R. A. Marcel, R. P. Miller, J. Material Science, Vol. 5, (1970), pp. 171.

[17] J. Frenkel, Kinetic Theory of Liquids, Oxford University Press, (1946) pp. 424.

[18] X. Zhou and M. Yamane, "Effect of Heat-Treatment for Nucleation on the Crystallization of $MgO-Al_2O_3-SiO_2$ Glass Containing TiO_2", J. Ceram. Soc. of Jpn, Vol. 96, No. 2, (1988), pp. 152-588.

[19] J. H. Jean and T. K. Gupta, "Crystallization kinetics of binary borosilicate glass composite", J. Mater. Res., Vol. 7, No. 11, (1992), pp. 3103-3111.

[20] J. H. Jean and T. K. Gupta, "Devitrification inhibitors in borosilicate glass and binary borosilicate glass composite", J. Mater. Res., Vol. 10, No.5, (1995), pp. 1312-1320.

[21] Y. Imanaka, N. Kamehara, and K. Niwa, "The Sintering Process of Glass/Alumina Composites", J. Ceram. Soc. of Jpn, Vol. 98, No. 8, (1990), pp. 812-816.

[22] H. Jebsen-Marwedel, Glastechn. Ber., Vol. 20, (1942) pp. 221.

[23] J. Loeffler, Glastechn. Ber., Vol. 23, (1950), pp. 11.

[24] H. Jebsen-Marwedel, Glastechn. Ber., Vol. 25, (1952), pp. 119.

[25] J. Widtmann, Glastechn. Ber., Vol. 29, (1956), pp. 37.

[26] Y. Imanaka, S. Aoki, N. Kamehara, and K. Niwa, "Crystallization of Low Temperature Fired Glass/Ceramic Composite", J. Ceram. Soc. of Jpn, Vol. 95, No. 11, (1987), pp. 1119-1121.

[27] Y. Imanaka, K. Yamazaki, S. Aoki, N. Kamehara, and K. Niwa, "Effect of Alumina Addition on Crystallization of Borosilicate Glass", J. Ceram. Soc. of Jpn, Vol. 97, No. 3, (1989), pp. 309-313.

[28] S. Nishigaki, S. Yano, J. Fukuta, M. Fukaya and T. Fuwa, "A NEW MULTILAYERED, LOW-TEMPERATURE-FIREABLE CERAMIC SUBSTRATE", Proceedings '85 International Symposium of Hybrid Microelectronics (ISHM), (1985), pp. 225-34.

[29] S. Wahisa, The Journal of Electronics and Communication Engineers of Japan, Vol 49, No. 7, (1966), pp. 37-47.

[30] N. Yamaoka, M. Masaru and M. Fukui, Am. Ceram. Soc. Bull., Vol. 62, No. 6, (1983) pp. 698.

[31] W. D. Kingery, H. K. Bowen and D. R. Uhlmann: Introduction to Ceramics (John Wiley & Sons, Inc., 1976).

[32] P. M. Sutton, "Dielectric Properties of Glass", Briks & Schulman Progress in Dielectrics II, (1960) pp. 114-161.

[33] V. Hippel, "Dielectric Materials and Its Application"

[34] V. D. Frechette, "Non-Crystalline Solids", (1960), pp. 412.

[35] Y. Imanaka, "Material Technology of LTCC for High Frequency Application", Material Integration, Vol. 15, No. 12 (2002), pp. 44-48.

[36] R. R. Tummala and A. L. Friedgerg, "Composites, Carbides-Thermal Expansion of Composite Materials", J. Appl. Phys., Vol. 41, No. 13, (1970), pp. 5104-5107.

[37] P. S. Turner, J. Res. NBS, Vol. 37, (1946), pp. 239.

[38] H. M. Kraner, Phase Diagrams, Material Science and Technology, 6-II, (1970), pp. 83-87.

[39] C. N. Fenner, J. Am. Ceram. Soc., Vol. 36, (1913), pp. 331-384.

[40] Y. Imanaka, S. Aoki, N. Kamehara, and K. Niwa, "Cristobalite Phase Formation in Glass/Ceramic Composites", J. Am. Ceram. Soc. Vol. 78, No. 5, (1995), pp. 1265-1271.

[41] Y. Imanaka, "Multilayer Ceramic Substrate, Subject and Solution of Manufacturing Process of Ceramics for Microwave Electronic Component", Technical Information Institute, (2002), pp. 235-249.

[42] Ryskewitsch, "Compression Strength of Porous Sintered Alumina and Zirconia", J. Am. Ceram. Soc., Vol. 36, No. 2, (1953), pp. 65-68.

[43] E. Orowan, "Fracture and Strength of Solids [Metals]", Repts. Progr. in Phys., Vol. 12, (1949), pp. 185-232.

[44] N. J. Petch, "Cleavage Strength of Polycrystals", J. Iron Steel Inst. (London), 174, Part I, May, (1953), pp. 25-28.

[45] S. Kistler, J. Am. Ceram. Soc., Vol. 45, No. 2, (1962), pp. 59.

[46] M. Nordberg, E. Mochel, H. Garfinkel, J. Olcott, J. Am. Ceram. Soc., Vol. 47, (1964) pp. 215.

[47] W. Weibull, "A Statistical Distribution Function of Wide Applicability", J. Appl. Mech. Vol. 18, (1951), pp. 293.

[48] R. M. Jones, Mechanics of Composite Materials, McGraw Hill, New York (1975).

[49] R. W. Davidge and G. Tappin, "Thermal Shock and Fracture in Ceramics", Trans. Br. Ceram. Soc. Vol. 66, (1967) pp. 405.

[50] D. P. H. Hasselman, Unified Theory of Thermal Shock Fracture Initiation and Crack Propagation in Brittle Ceramics, J. Am. Ceram. Soc. Vol. 49, (1969), pp. 68.

[51] W. D. Kingery, H. K. Bowen and D. R. Uhlmann: Introduction to Ceramics 2nd ed. (John Wiley & Sons, Inc., (1976), pp. 634.

Chapter 3

Conducting material

3.1 Introduction

With LTCCs, conductors are achieved by patterning conductive paste onto a ceramic green sheet by screen printing, then firing it simultaneously with the ceramic. The conductive material is one of the important constituent materials of LTCCs, and many characteristics are required of it so that it can meet a wide range of applications in high frequency components, wiring for substrates, and electrode terminals.

Firstly, with the purpose of reducing transmission loss of high speed, high frequency signals, it is necessary to use material with low electrical resistance for the conductive material. As shown in Table 2-1 in the previous chapter, among the various metals Cu, Au, Ag and metals alloyed with them have low electrical resistance, and they can be considered appropriate for wiring in high frequency substrates.

Furthermore, in order to achieve a monolithic module by simultaneously firing the conductor layer metal and dielectric layer ceramic, as shown in Figure 3-1 [1, 2], it is important that the optimal firing temperature of both materials nearly matches, and there is a match between the firing shrinkage behavior of both materials. As well as controlling the shrinkage behavior between both materials, it is essential to form good adherence between both materials with the aim of ensuring electrical and mechanical reliability of the LTCC itself by taking the conductor/dielectric interfacial phenomenon into consideration.

When the LTCC maintains the electronic devices mounted on it with a voltage applied for a long time in conditions of high temperature and high humidity, if the circuit design and material design are inappropriate, the migration phenomenon may occur in the conductor, and failures such as insulation failure, shorting and the like may occur. The impact on the material factors of the conductor is significant, and there is a requirement to prevent migration of the conductor with an appropriate approach to the conductor material, and also to improve the migration resistance of these materials.

In addition, the LTCC surface metallized electrode layer is frequently required to connect the active and passive components electrically and mechanically. Soldered joints and wire bonding are the representative connection methods, and it is necessary to select conductive materials suited to either method.

When LTCCs are used as substrates for LSI high density packaging, they must release the heat generated by the chips efficiently, and since ceramic dielectrics have low thermal conductivity, special thermal conduction paths called thermal vias are often formed in the LTCC. In this case, high thermal conductivity metals with excellent heat transference are required as conductive materials.

While the proportion of the conductive material itself in the LTCC is small at only several vol%, during the process, the residual portion of the conductor formed by screen printing using a paste of metal powder dispersed in a vehicle adheres to the screen and the squeegee and it is lost during cleaning. The lost amounts are often greater than expected and in order to achieve low cost LTCCs, it is necessary to keep down the price of the metal powder used in the conductive paste. From this point of view, of those metals suited for use as LTCC conductive materials, Au bullion is expensive and is probably impractical due to its price.

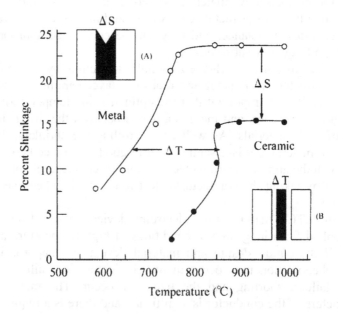

Figure 3-1   The firing shrinkage behavior of ceramic material and conductive materials in LTCCs [Ref. 1].

The sections below summarize firstly the metalization of conductive materials on alumina ceramics (thick film processing and HTCC) which has a strong connection with the metalization of LTCC conductive materials, and further explain in detail the requirements of the conductive materials above.

3.2 Conductive paste materials

Conductive paste is obtained by dispersing conductive metal powder (conductive material) in an organic vehicle as a binder or plasticizer using a roll mill, and kneading it in a mixer. Metal powder is used as the conductive ingredient and for inorganic additives, normally low melting point glass and reactive oxides are used. The organic vehicle is composed of retarder thinners such as terpineol, texanol and the like, organic binders (cellulose, acrylic, butyral resin and so on), plasticizers, dispersing agents, thixotropic agents and so on. Figure 3-2 shows the sheet resistance after firing of the different kinds of metal paste used in thick film processing. In general, conductors include inorganic additives, and since they are obtained by sintering powder, they form voids readily inside the conductor and generally they show higher resistance than the specific resistance of the metal itself. Incidentally, sheet resistance ($\Omega/\square$) is the bulk resistance ($\Omega \cdot$ cm) of the conductive material divided by the nominal thickness (normally 0.0025 cm), and it is frequently used as a unit of resistance for thick film conductors.

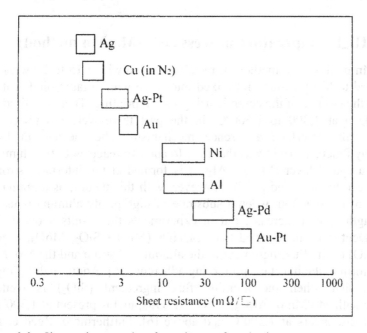

Figure 3-2　Sheet resistance values for each type of conductive paste.

3.3   Metallization methods for alumina ceramics

Metallization methods for alumina ceramics using conductive paste can be broadly classified into two types; thick film processing and cofiring (HTCC). Although both methods share many points of similarity in the printing process, big differences can be seen in the other manufacturing processes.

Below, we consider alumina substrates, developed in advance of LTCC metallization technology, as the basis for thick film metallization and metallization for HTCC, and establish the differences from LTCC metallization.

3.3.1   Thick film metallization

With thick film processing, conductors are generally fabricated as follows. By applying pressure with a squeegee to conductive paste containing metal powder, a film is applied to the substrate, normally through the openings in a screen made of stainless steel mesh. Next, the film applied with printing to the substrate is fired, in order to achieve densification of the conductor film and adherence with the substrate. With this method, production of films with thicknesses of 1 to 30 µm, and line widths of more than 50 µm is possible.

As shown below, thick film metallization for alumina substrates has two further types, a high temperature process type and low temperature process type.

3.3.1.1   High temperature process type (Mo-Mn method)

The Mo-Mn method is a method of metallization in which 15 to 20 wt% of Mn is added to Mo powder, and mixed into paste with organic binder, it is printed on the surface of the ceramic using screen printing. Then it is fired in wet hydrogen at 1,300 to 1,500°C. In the past there were a number of theories of the interface adherence mechanism. The mechanism first advanced by Pincus in 1953 was that the Mo and Mn reacts with the alumina, and since a spinel layer (MnO • Al_2O_3) is formed at the interface, strong adherence can be achieved [3, 4]. However, with this theory, as alumina is the source of the reaction, when a substrate of high purity alumina is used, bond strength should increase, but in experiments the results showed the opposite. Denton et al. reported that reactants (MnO • SiO_2, MoO_3, MgO, MoO_3 • CaO) of the flux ingredient in the alumina substrate and the Mo, Mn and so on form at the interface promoting adherence [5]. Additionally, Floyd established that when the alumina flux ingredient (SiO_2) is wetted thoroughly with Mo/Mn or Al_2O_3, MnO • Al_2O_3 spinel is present at 1,300°C, although it disappears at 1,500°C and above [6]. Furthermore, Reed et al.

found that MnO • Al$_2$O$_3$ spinel melts and disappears in alumina flux glass phase at high temperatures [7].

As a result of the research described above, the bonding mechanism of the Mo-Mn method can be thought of as follows [8, 9, 10, 11]. (1) The Mn in the paste is oxidized by the wet gas to become MnO. (2) The MnO reacts with the alumina, forming spinel. Next it reacts with the flux in the alumina and forms a high fluidity glass phase (afterwards, the spinel may disappear). (3) The glass formed penetrates the voids in the Mo metallization layer by the capillary phenomenon. (4) This glass then hardens making an anchor between the substrate and metallization layer, forming a strong bond.

Since the flux in the alumina substrate promotes migration of the interface reactants, MnO is added to the paste when using high purity alumina substrates [12]. Furthermore, in order to increase the reliability of the bond, in some cases an alumina substrate with added MnO (called pink grade because it is pink) is used first.

3.3.1.2 Low temperature process type

In the low temperature process type, Cu, Ag, Au or alloys of these metals are used as conductive ingredients, and firing is performed at around 900°C, below the melting point of each metal ingredient. Types of bonding are divided into frit bonding with the glass ingredients in the alumina substrate; chemical bonding with adherence achieved due to chemical reaction with the alumina; and mixed bonding that is a combination of both of these [13].

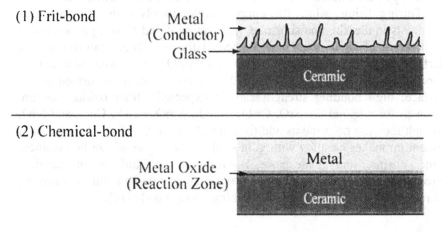

Figure 3-3   Schema of low temperature type thick film metallization of alumina substrates (1) Frit bond type and (2) chemical bond type.

(1) Frit bonding

With this type of paste, glass frit is mixed in conductive metal powder as an addition agent for promoting adherence with the substrate. Glass containing PbO or Bi_2O_3 is used for this frit material. Since the softening point of this glass is low, it penetrates the grain boundaries in the alumina substrate, fusing them with the glass ingredients in the alumina substrate, interlocking and bonding [14]. Kinetic analysis of penetration of glass containing PbO and Bi_2O_3 in the alumina substrate is under investigation, and it has been established that by increasing the amount of PbO and Bi_2O_3 added, the viscosity of the glass falls and with capillary force as the driving force, the penetration depth in the alumina substrate is increased. As a result of research regarding interface reactions of $PbO-SiO_2$ type glass and alumina substrates, it has been confirmed that glass penetrates the alumina grain boundary phase, and the width of the grain boundary phase expands [15]. In addition, in glass with ZnO added, compounds containing ZnO and Al_2O_3 are formed (ex. spinel phase $ZnO \cdot Al_2O_3$). In general for frit for atmospheric firing, the $PbO-SiO_2-B_2O_3$, $PbO-SiO_2-ZnO$ and $PbO-ZnO-B_2O_3$ families are used [16]. In paste containing copper powder for firing in a nitrogen atmosphere, glass with a composition that does not deoxidize readily (SiO_2-$CaO-Al_2O_3$ family) is used [17]. There is also a tendency recently to use lead-free glass out of concern for environmental issues.

(2) Chemical bonding

As inorganic additives when performing atmospheric firing with conductive paste in which gold or silver alloy is the main ingredient, the following several types are known (Cu, Cu_2O or CuO, GeO_2, ZnO, CdO, and Bi_2O_3) [18]. During firing, while the copper oxide dissolves in the glass flux ingredients in the alumina substrate, penetrating it and forming an anchor, a spinel phase of $CuO \cdot Al_2O_3$ is formed at the interface, achieving strong interface adherence. Similarly ZnO and CdO react with alumina at temperatures lower than 1,000°C, and as a spinel phase is formed at the interface, high bonding strength can be expected. Other oxides that are known to form spinel are SrO, Cr_2O_3, MoO_3, MnO, Fe_2O_3, CoO, and NiO, although they are not used as additives for chemical bonding. During firing, germanium makes an alloy with gold so that a strong bond can be obtained. In the substrate and in the paste, Bi_2O_3 promotes liquidation of the glass ingredients, and since it achieves a strong anchor with the alumina substrate, it is regarded as being effective for making strong bonds [19].

3.3.2 Cofiring metallization

This is achieved by screen printing conductive paste on ceramic green sheets (raw sheets of ceramic powder and organic binder) before firing, forming electrode patterns, superimposing these printed green sheets, and laminating a monolithic ceramic green body with thermocompression bonding. Next this body is fired at high temperature, simultaneously densifying the ceramic and forming a metallization layer [20, 21, 22, 23]. In order to fire the conductor simultaneously with the ceramic, it is necessary to consider the firing temperature (melting point of the metal), the firing environment (oxidation-reduction potential of the metal), reactions between the ceramic and metallization layer and so on. Therefore, the conductor metallization materials are limited by the kinds of ceramics used. The differences from the manufacturing processes of thick film processing can be seen below.

(1) Printing process
In thick film processing, conductive paste is printed on fired ceramic substrates, however in cofiring the printing is on ceramic green sheets. Ceramic green sheets are flexible sheets of ceramic powder and organic binder with porosity of around 40%. When printing on green sheets, the solvent in the conductive paste penetrates the green sheet and dissolves the resin in the green sheet. Although in extreme cases, phenomena such as bleeding of the pattern and the like can be seen, since it allows the conductor to make an anchor in the ceramic before firing, it is beneficial for achieving high adhesion after firing.

(2) Laminating process
With cofiring, in the laminating process heat and pressure are applied to the conductor printed on the green sheet (a process not used for thick film processing). As a result, it is possible to raise the packing density of the metal powder in the printed conductor. Furthermore, the metal powder and binder resin fluidize with heating, and with this fluidity the ceramic and metal powder intermigrate at the interface, forming a mechanical anchor.

(3) Firing process
With thick film processing, there is no dimensional change of the ceramic substrate during firing, however with cofiring the ceramic and metal both contract as they are sintered during firing. For this reason, while conductor films that are cofired are dense, conductor films made with thick film processing are frequently porous. This is because with cofiring, the ceramic base itself contracts, and the conductor film has freedom to contract in the x and y axes in addition to the z axis, allowing it to densify readily. On the other hand, with thick film processing, contraction of the film is restricted in the x and y axes, and since it occurs only in the z axis, densification is inhibited.

3.3.2.1 Metallization for HTCC

When alumina is used as the ceramic for cofiring (HTCC), the conductive materials are Mo and W (both materials have half the value of resistance of Pt and Pd, have excellent adherence, and are inexpensive). Even when either Mo or W are used as conductive materials, firstly, the glass of the sintering additive added to the alumina at around 1,450°C begins to soften and flow. Then this glass flows into the porous metal conductor during sintering, forming a mechanical anchor (the possibility of chemical reaction between the glass phase and Mo and W has been suggested, but in fact the formation of reactants has not been reported), thus forming the metallization layer. The wetting of the sintering additive (SiO_2-MgO-Al_2O_3 glass) in the alumina and the tungsten is the key to the bonding of the metallization, and it has been reported that in order to achieve good fluidity of the glass ingredients and high bonding strength, alumina ceramic with a relatively high percentage of glass of around 8% is optimal [24, 25, 26]. In addition, Wilcox et al achieved a compound of Mo + Cu by injecting Cu by the infiltration method into the porous Mo conductor (Mo and Cu do not react), establishing that the value of resistance of the conductor can be lowered [27].

Wet hydrogen with adjusted oxygen partial pressure is being used in order to prevent oxidation of the Mo and W in the firing environment when firing [28].

3.4 Conductivity

Au, Ag, and Cu, which have low electrical resistance, are appropriate as conductors for use in LTCCs. Table 3-1 shows the value of resistance of Cu and Au when various impurities are dissolved at 1 at .%. The value of resistance of the main material changes markedly depending on the impurity element. For example, if Ti is dissolved at 1 at .% in Cu, the value of resistance is 10 times that of pure copper [29, 30, 31, 32, 33]. Furthermore, if the copper reacts with the other metal and an intermetallic compound is formed, it is in most cases more than 10 $\mu\Omega$ • cm. The following compounds are less than 10 $\mu\Omega$ • cm [34, 35].

Al_2Cu: 6 $\mu\Omega$ • cm, CuAl: 3 $\mu\Omega$ • cm, Cu_3Au: 4 $\mu\Omega$ • cm, CuZn: 5 $\mu\Omega$ • cm, Cu_3Ge: 5.$\mu\Omega$ • cm, Cu_2Mg: 8 $\mu\Omega$ • cm, $CuMg_2$: 9 $\mu\Omega$ • cm

Therefore, in order to achieve high conductivity it is necessary to avoid contamination from impurities at all costs, and to maintain high purity. Figure 3-4 shows the variation in value of resistance when Pd and Au are dissolved in Ag (both Ag-Pd and Ag-Au types are solid solution alloys). As noted below, in the case of Ag conductors, there have been many attempts to make alloys in order to improve anti-migration characteristics, but as shown in the figure, with the Ag-Pd type for example, when Pd is dissolved at 60 wt%, the value of resistance rises to 40 times that of pure silver. If Pd is added to increase migration resistance, but the value of resistance rises, the point of using Ag is lost. As noted above, it is necessary to understand fully the impact of the added element and composition on electrical resistance.

Table 3-1   The value of resistance of Cu and Au when impurities are dissolved at 1 at .%.

Impurity element	Cu (1.7 μΩ • cm)	Au (2.1 μΩ • cm)
Ag	1.9	2.8
Cu	-	3.6
Au	2.5	-
Pd	2.6	2.9
Al	2.7	4.0
Ni	2.8	5.1
Pt	3.7	3.3
Sn	4.8	7.6
Co	8.6	7.9
Ti	17.5	15.0

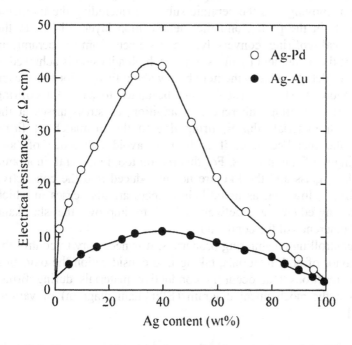

Figure 3-4   The composition dependence of resistance values of Ag-Pd and Ag-Au type alloys [Both Ag-Pd and Ag-Au types are solid solution type alloys].

3.5 Suitability for co-sintering

Figure 3-1 is a conceptual diagram of the effect of mismatch between the firing shrinkage rate of the metal and ceramic materials. In the figure, the difference in temperature for the start of firing shrinkage for both materials is shown as ΔT, while the difference in final firing shrinkage when sintering is complete is shown as ΔS. In this figure, if for example the temperature at which sintering of the ceramic is complete is 850°C, the temperature at which sintering of the metal is complete is 600°C. The difference ΔS is caused by the formation of cavity-like areas inside the substrate and on the surface of the conductor where the firing density is uneven. As a result, the ceramic substrate becomes distorted, and the dimensional accuracy of the wiring is hard to control. ΔT can be considered a cause of adherence defects in the conductor/ceramic interface. In order to reduce the mismatch in firing shrinkage, the particle size of the conductive material, and of its composition and additives are optimized. By adjusting the ceramic material parameters, approaches are simultaneously being made to reduce the amount of mismatch. Unlike thick film processing, as the sintering of cofired conductors is facilitated by interaction with the ceramic, the conductive paste used for LTCCs often consists of only the conductive metal powder and organic vehicle, and other inorganic substances are not added (refer to Figure 3-5). In the thick film processing of the low temperature process type, the paste begins moving into the ceramic substrate (including the inorganic glass ingredients in the paste), and the metallization layer adheres to the substrate. But with cofiring, conversely, the force acts from the ceramic in the direction of the metal, and by mass transfer, high adhesion is achieved at the interface (for details, refer to the item below about firing). For this reason, inorganic components should probably not be added to paste for cofiring with the aim of promoting adherence. In addition, as stress arises at the interface in both materials during firing due to the mismatch of firing shrinkage of the metal/ceramic, it is best to avoid adherence of both materials during the firing process. For this reason too, in general, inorganic additives such as glass and the like are not introduced into the conductive paste for cofiring. However, as noted below, there are also cases in which some ceramic ingredients are actively added to improve the shrinkage behavior and adherence of the conductor.

When controlling co-sintering properties, it is necessary to control the sintering behavior of both materials, taking into consideration the oxidation and reductive reactions that occur in conductive materials during firing. Figure 3-6 is a thermochemical diagram (Ellingham diagram) of various types of metal.

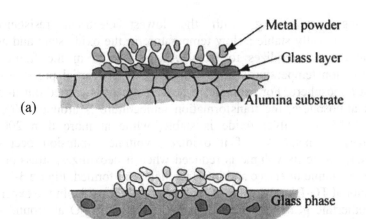

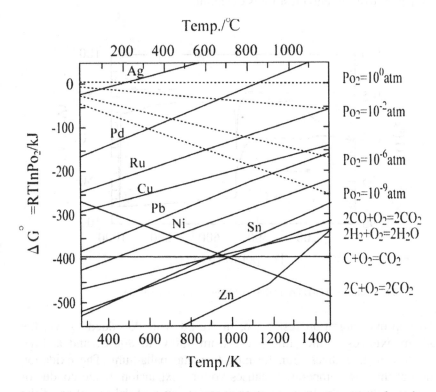

Figure 3-5   The differences in the microstructures of a thick film processing conductor formed on an alumina substrate and an LTCC conductor.

Figure 3-6   Thermochemical diagram (Ellingham diagram) of various types of metal.

Among metals, Ag, with the lowest electrical resistance, is thermodynamically stable at low temperature in the oxide state and at high temperature, it deoxidizes to metal Ag, as shown in the figure. This transformation temperature varies with the oxygen partial pressure of the ambient atmosphere. For example, if heat treatment is carried out in a pure oxygen atmosphere, the transformation temperature is around 200°C, and below 200°C the silver oxide is stable, while at more than 200°C, it deoxidizes to metal Ag. If it oxidizes, volume expansion occurs and conversely, since its volume is reduced when it deoxidizes, stress arises at the metal/ceramic interface and interface flaws are formed. Figure 3-7 shows the results of TG-DTA analysis of high purity silver oxide. In the experiment, an endothermic peak indicating deoxidation of the Ag_2O at around 480°C was seen, at a higher temperature by some 300°C than expected from the thermodynamic diagram, with weight loss of around 7% [36]. The comparatively low speed of the reductive reaction can be considered to have an impact on the difference between the thermodynamic prediction and the experimental data, and according to the heat profile, the apparent reduction temperature rises and falls easily. In order to control the interfacial phenomenon, experimental testing is essential.

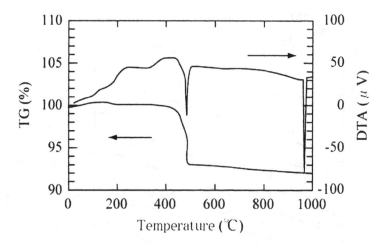

Figure 3-7   The results of TG-DTA analysis of high purity silver oxide.

To improve migration resistance, in Ag-Pd with Pd added to Ag, the palladium oxidizes in the atmosphere from 450°C to 800°C, and at high temperature it deoxidizes, transforming to metal palladium. The oxidation reaction at this low temperature causes volume expansion of the conductor element, and during the firing process, it is the cause of delamination of the multilayer ceramic [37]. In the same Ag-Pd composition, the degree of oxidation of the Pd differs [38] depending on the production method. Since the oxidation reaction of palladium in powder produced with the molten alloy method is suppressed compared with that made using the coprecipitation method and simply mixed Ag-Pd powder, it is effective in preventing delamination. Figure 3-8 shows the results of X-ray analysis of Ag-Pd powder produced using the various methods. In the type where Ag powder and Pd powder are simply mixed, the separate peaks of Ag and Pd are observed. In the molten alloy method where Ag-Pd are melted, fully dissolved and alloyed, the peaks of the solid solution are seen. The Ag-Pd produced with the coprecipitation method shows a structure that is between that produced by the molten alloy method and simple mixing method. Incidentally, the coprecipitation method is a technique in which nucleus formation and growth is performed from a solution state, then solids are precipitated. In the case of Ag-Pd, by deoxidizing the nitric acid solution of the Ag-Pd with hydrazine, Ag-Pd powder is precipitated at around 250°C.

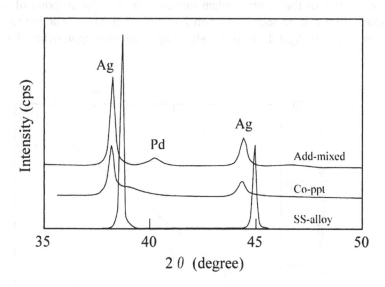

Figure 3-8 The results of X-ray analysis of Ag-Pd powder produced using various methods (simple mixing, coprecipitation, solid solution alloy method).

Furthermore, when using Cu as a conductive material, the fundamental issues are the same, although Cu oxidation and deoxidation behavior, and the various problems that arise when cofiring, are covered in detail in the section on firing.

3.6 Adherence

Adherence between the conductor and ceramic is achieved due to the glass ingredients in the ceramic penetrating the conductor during cofiring, and forming a mechanical anchor. For this reason, we noted that secondary inorganic components are not added to the conductive material. However, to adjust for better adherence and firing shrinkage, it is very effective to mix into the conductive paste a 'companion powder' of ceramic that is one component of the ceramic in the conductive paste or that has the same composition. Figure 3-9 shows the conductor adherence strength and firing shrinkage behavior of an Ag-Pd conductor when ceramic (glass/ceramic composite) is added to the conductor. The companion powder added to the conductor combines with the ceramic during firing, forming a strong interlock (refer to Figure 3-10), and achieving a strengthening effect. Furthermore, since the ceramic is sintered in the conductor when ceramic is added to the conductor, it is possible to bring the firing shrinkage behavior closer to that of the ceramic when looked at from a macro point of view. In Figure 3-9 it can be seen that when 20 vol% of ceramic is added to a 60/40 composition of Ag-Pd, its firing shrinkage behavior approaches that of the ceramic.

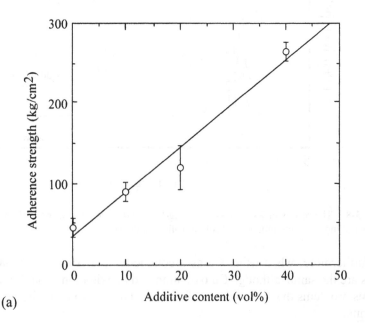

(a)

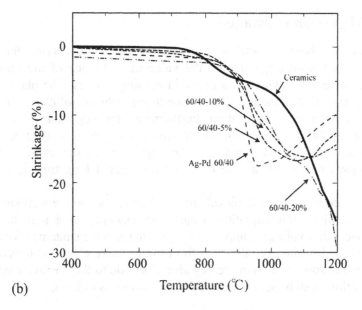

Figure 3-9   Conductor adherence strength (a) and firing shrinkage behavior (b) of an Ag-Pd conductor when ceramic (glass/ceramic) is added to the conductor.

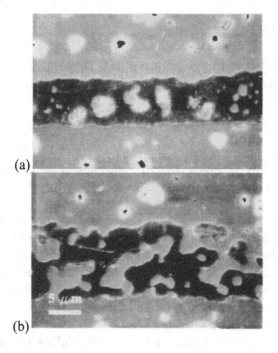

Figure 3-10   Microstructure of a Ag-Pd conductor with ceramic added at (a) 10 vol% and (b) 40 vol%.

3.7 Migration resistance

Migration is where the metal used as wiring and electrodes moves the top or inside of the insulator, reducing the value of resistance of the insulation between electrodes, which is a factor in causing failures. The phenomenon causing migration of metal elements due to the influence of electric fields is given the name electromigration. Furthermore, the migration of the metal elements is the reason for the electrolytic phenomenon of H_2O and what results is known as ion migration. The migration that occurs in electronic components and so on as the result of use over a long time is this ion migration.

Silver, copper, tin, lead, nickel, gold, solder and so on are well known as metals in which ion migration occurs. For example, in a high humidity atmosphere, if a voltage is applied to Ag wiring and is maintained for a long time, silver dendrite crystals grow from the negative electrode towards the positive electrode, and from the negative electrode to the positive electrode, Ag_2O colloids can be seen. This mechanism works as follows.

1) Ionization occurs due to the potential difference between the Ag electrodes and the presence of absorbed water in the surrounding atmosphere.
$$Ag \rightarrow Ag^+$$
$$H_2O \rightarrow H^+ + OH^-$$

2) Ag^+ and OH^- form and precipitate AgOH at the positive electrode.
3) The AgOH decomposes and becomes Ag_2O at the positive electrode and disperses in colloids.
$$2AgOH \longleftrightarrow Ag_2O + H_2O$$

4) Next, in a hydration reaction, the reaction $Ag_2O + H_2O \longleftrightarrow 2AgOH \longleftrightarrow 2Ag^+ + 2OH^-$ proceeds, and Ag+ migrates to the negative electrode, and precipitation of Ag dendrite crystals proceeds.

Furthermore, ion migration of the copper also occurs when, due to the presence of electric fields and water, the reaction $Cu + 4H_2O \rightarrow Cu(OH)_2 + O_2 + 3H_2$ occurs at the positive electrode. On the other hand, the reaction $Cu(OH)_2 \rightarrow CuO + H_2O$ proceeds at the negative electrode and CuO dendrites grow from the negative electrode to the positive electrode.

Additionally, an anion X^{n-} (for example, Cl^-) is involved,
$$Cu_2O + 2/n\, X^{n-} + H^+ \rightarrow CuX_{2/n} + H_2O + Cu$$
$$Cu + n\, X^- \rightarrow CuX_n + ne^-, \quad CuX_n \rightarrow Cu^{n+} + nX^-, \quad Cu^{n+} + ne^- \rightarrow Cu$$
By the above reaction, metal Cu is precipitated.

The time required for dielectric breakdown due to the migration phenomenon can be calculated and predicted with an empirical formula with temperature and humidity as parameters.

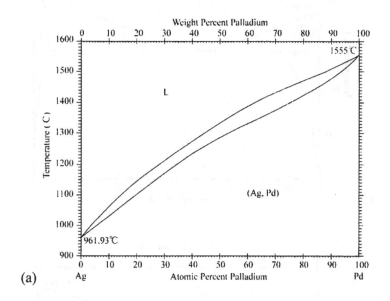

(a)

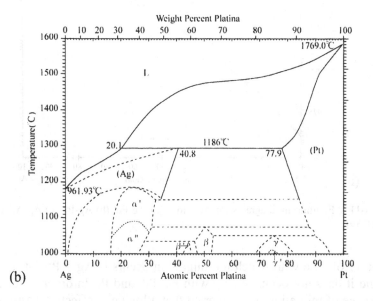

(b)

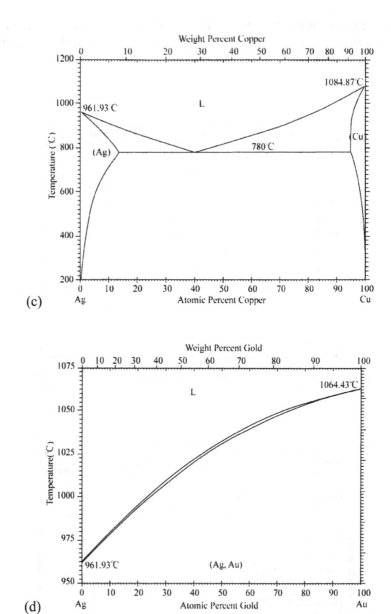

Figure 3-11   Equilibrium diagrams for Ag alloys (a) Ag-Pd, (b) Ag-Pt, (c) Ag-Au, (d) Ag-Au [Ref. 39].

The order of ease with which ion migration occurs is Ag > Pb > Cu > Sn > Au, and it does not occur readily with Fe, Pd, and Pt. In order to retard the occurrence of migration, it is reported that alloying is effective (Ag-Pd, Ag-Pt, Ag-Cu, Ag-Au in the case of Ag). As is clear from the equilibrium diagrams in Figure 3-11 [39], Ag-Pd, Ag-Pt, and Ag-Au are solid solution alloys and Ag-Cu is a eutectic alloy. In the solid solution types, at the atomic level, and in the eutectic type at the micro level, suppression of migration of Ag is thought to be a factor in improving migration resistance. However, since the melting point changes with alloying as shown in Figure 3-4, it is necessary to give careful consideration when deciding the alloy composition so as to avoid raising its value of electrical resistance. The general tendency of composition dependence with regard to characteristics such as electrical resistance and so on of each alloy type is shown in Figure 3-12.

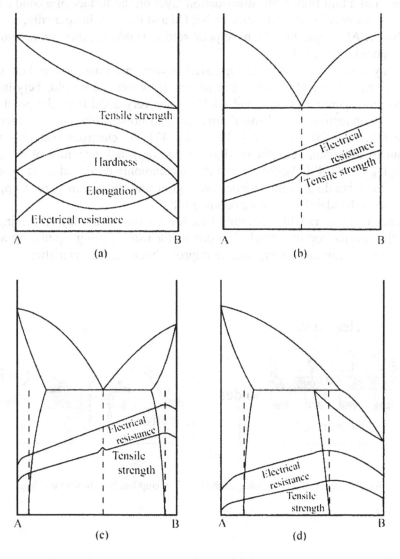

Figure 3-12   The composition dependence of various characteristics of each type of alloy (a) solid solution, (b) eutectic (not dissolved), (c) eutectic, (d) peritectic.

3.8   Bondability (solder wetability and wire bondability)

When mounting various active and passive components on LTCCs, soldered joints or wire bonding are generally used (refer to Figure 3-13). Ag, Cu, Ni, Au, Pd, Pt and their alloys have excellent solder wetability with the widely used lead eutectic solder (63Sn/37Pb), and they are used as the electrode material for soldered joint pads [40]. Although Au has very good solder wetability, since it reacts with the solder and embrittles the joint surface, it is not used for the joint electrode material itself. However, it is often used in the form of a thin film as an antioxidation layer on the surface of a conductor with good solder bondability such as Ni, Cu and the like. Incidentally, as Fe, Cr, Ni-Cr, Al, Ti and the like have poor solder wetability, they are not good for soldered joints [41].

Many materials [42, 43] are suggested as wire materials for wire bonding such as Mg-Si, Al-Mg, Al-Cu and so on. However, pure gold, beryllium doped gold, pure aluminum, and Al-1%Si are very good from the point of view of conductivity, mechanical strength, and bondability, and in general, these four materials are used [44, 45, 46, 47]. As electrode materials for terminals, aluminum, aluminum alloy (including small quantities of Si, Cu, or Mg), gold, Au-Pd, Ag-Pd, Ni and Pt are commonly used, and these metals show good bonding characteristics with gold and aluminum wire. Copper has inferior bondability for wire bonding [48].

Since conductors with a bonding function are the surface layer, forming a thin film on the top of a thick film conductor using plating, sputtering and the like is a technique widely used to improve bonding characteristics.

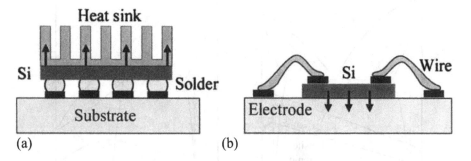

Figure 3-13   Typical LSI chip bonding methods a) Flip chip bonding, b) wire bonding.

References

[1] R. R. Tummala, "Ceramic and Glass-Ceramic Packaging in the 1990s, " J. Am. Ceram. Soc., Vol. 74, No. 5, (1991), pp. 895-908.

[2] Y. Imanaka and N. Kamehara, "Influence of Shrinkage Mismatch between Copper and Ceramics on Dimensional Control of the Multilayer Ceramic Circuit Board, " J. Ceram. Soc. Jpn., Vol. 100, No. 4, (1992), pp. 560-564.

[3] A. G. Pincus, "Metallographic Examination of Ceramic-Metal Seals," J. Am. Ceram. Soc., Vol. 36, No. 5, (1953), pp. 152-158.

[4] A. G. Pincus, "Mechanism of Ceramic to Metal Adherence of Molybdenum to Alumina Ceramics," Ceram. Age, Vol. 63, No. 3, (1954), pp. 16-20, pp. 30-32.

[5] E. P. Denton, H. Rawson, "The Metallizing of High-Al_2O_3 Ceramics," Trans. Brit. Ceram. Soc., Vol. 59, (1960), pp. 25.

[6] J. R. Floyd, "Effect of Composition and Crystal Size of Alumina Ceramics on Metal-to-Ceramic Bond Strength, " Ceramic Bull. Vol. 42, No. 2, (1963), pp. 65-70.

[7] L. Reed, R. A. Hnggins, "Electron Probe Microanalysis of Ceramic-to-Metal Seals," J. Am. Ceram. Soc., Vol. 48, No. 8, (1965), pp. 421-426.

[8] J. T. Klomp, "Interfacial Reactions Between Metals and Oxides during Sealing, " Ceramic Bull., Vol. 59, No. 8, (1980), pp. 794-799.

[9] K. White and D. P. Kramer, "Microstructure and Seal Strength Relation in the Molybdenum-Manganese Glass Metallization of Alumina Ceramics, " Materials Science and Engineering, 75, (1985), pp. 207-213.

[10] Y. S. Sun, J. C. Driscoll, "A New Hybrid Power Technique Utilizing a Direct Copper to Ceramic Bond", IEEE Transactions on Electron Devices, Vol. ED-23, No. 8, Aug., (1976), pp. 961-967.

[11] D. M. Mattox and H. D. Smith, "Role of Manganeses in the Metallization of High Alumina Ceramics, " Ceramic Bull., Vol. 64, No. 10, (1985), pp. 1363-1367.

[12] K. Otsuka, T. Usami, and M. Sekihara, "Interfacial Bond Strength in Alumina Ceramics Metallized and Cofired with Tungsten, " Ceramic Bull., Vol. 60, No. 5, (1981), pp. 540-545.

[13] M. L. Minges, Electronic Materials Handbook Volume 1 PACKAGING (ASM Internatioanl, 1989).

[14] Y. Kuromitsu, "Interfacial Reaction between Glass and Ceramics and the Application for AlN Circuit Board for Mounting Semiconductors, " Kyushu University Ph. D Thesis, 1997.

[15] Y. Kuromitsu, H. Yoshida, H. Takebe, and K. Morinaga, "Interaction between Alumina and Binary Glasses.", J. Am. Ceram. Soc., Vol. 80, No. 6, (1997), pp. 1583-87.

[16] K. Harano, K. Yajima, and T. Yamaguchi, J. Ceram. Soc. Jpn., Vol. 92, No. 9, (1984), pp. 504-509.

[17] R. W. Vest, "Materials Science of Thick Film Tecnology", Ceramic Bull., Vol. 65, No. 4, (1986), pp. 631-636.

[18] M. V. Coleman, G. E. Gurnett, "The Limitations of Reactively-Bonded Thick Film Gold Conductors", Solid State Technology, Mar.(1979), pp. 45-51.

[19] Y. Akimoto, "Bonding of Thick Film Electrode to Ceramics," Electro-Ceramics, Vol. 19, No. 11, (1988), pp. 54-60.

[20] W. G. Burger, C. W. Weigel, "Multi-Layer Ceramic Manufacturing", IBM J. Res. Develop., Vol. 27, No. 1 Jan. (1983), pp. 11-19.

[21] B. T. Clark and Y. M. Hill, "IBM Multichip Multilayer Ceramic Modules for LSI Chips-Design for Performance and Density", IEEE Transactions on Components, Hybrids, and Manufacturing Technology, Vol. CHMT-3, No. 1, March, (1980), pp. 89-93.

[22] N. Kamehara, Y. Imanaka, and K. Niwa, "Multilayer Ceramic Circuit Board with Copper Conductor", Denshi Tokyo, No. 26, (1987), pp. 143-148.

[23] Y. Imanaka, A. Tanaka, and K. Yamanaka, "Multilayer Ceramic Circuit Board –Wiring Material: From Copper To Superconductor-," FUJITSU, Vol. 39, No. 3, June (1988), pp. 137-143.

[24] R. R. Tummala, E. J. Rymaszewski:Microelectronics Packaging Handbook,VAN NOSTRAND REINHOLD (1989).

[25] G. Toda, T. Fujita and S. Ishuhara , "Sintering and Wetting – Sintering of Alumina Substrate, " Material Science, Vol. 3, No. 2, (1985), pp. 81-86.

[26] K. Otsuka, T. Usami, and M. Sekihara, "Interfacial Bond Strength in Alumina Ceramics Metallized and Cofired with Tungsten," Ceramic Bull., Vol. 60, No. 5, (1981), pp. 540-545.

[27] D. A. Chance and D. L. Wilcox, "Capilary-Infiltrated Conductors in Ceramics," METALLURGICAL TRANSACTIONS, Vol. 2, March, (1971), pp. 733-741.

[28] D. A. Chance, D. L. Wilcox: "Metal-Ceramic Constraints for Multilayer Electronic Packages", Proceedings of the IEEE, Vol. 59, No. 10 Oct. (1971), pp. 1455-1462.

[29] R. A. Matula, J. Phys. Chem. Ref. Data 8, (1979), pp. 1147.

[30] P. D. Desai, H. M. James, C. Y. Ho, J. Phys. Chem. Ref. Data 13, (1984), pp. 1131.

[31] G. K. White, S. B. Woods, Phil. Trans. Roy. Soc. London Ser. A251, (1958), pp. 273.

[32] T. E. Chi, J. Phys. Chem. Ref. Data 8, (1979), pp. 439.

[33] P. D. Desai, T. K. Chu, H. M. James, C. Y. Ho, J. Phys. Chem. Ref. Data 13, (1984), pp. 1069.

[34] Handbook of Electrical Resistivities of Binary Metallic Alloys (CRC Press, 1983).

[35] J. M. E. Harper, E. G. Colgan, C. K. Hu, J. P. Hummel, L. P. Buchwalter and C. E. Uzoh, "Material Issues in Copper Interconnections", MRS Bull./AUGUST, (1994), pp. 23-29.

[36] Y. Imanaka, "Material Technology of LTCC for High Frequency Application," Material Integration, Vol. 15, No. 12 (2002), pp. 44-48.

[37] S. S. Cole, "Oxidation and Reduction of Palladium in the Presence of Silver," J. Am. Ceram. Soc., Vol. 68, No. 4, (1985), pp. C106-C107.

[38] S. F. Wang, W. Huebner, and C. Y. Huang, "Correlation of Subsolidus Phase Relations in the Ag-Pd-O System to Oxidation/Reduction Kinetics and Dilatometric Behavior," J. Am. Ceram. Soc., Vol. 75, No. 8, (1992), pp. 2232-39.

[39] T. B. Massalski, Binary Alloy Phase Diagram 2nd Ed., ASM International, 1990.

[40] 13) J. M. E. Harper, E. G. Colgan, C-K. Hu, J. P. Hummel, L.P. Buchwalter, and C. E. Uzoh, "Materials Issues in Copper Interconnections," MRS Bull., Vol. XIX, No. 8, August (1994), pp. 23-29.

[41] D. P. Seraphim, R. C. Lasky and C-Y Li: Principles of Electronic Packaging, (McGraw-Hill, Inc., 1989), pp. 594.

[42] T. J. Matcovich, 31st Proceedings of Electronic Components Conference, (1981) pp. 24-30.

[43] J. Onuki, M. Suwa, and T. Iizuka, 34th Proceedings of Electronic Components Conference, (1984), pp. 7-12.

[44] S. P. Hannula, J. Wanagel, and C. Y. Li, 33rd Proceedings of Electronic Components Conference, (1983), pp. 181-188.

[45] M. Poonawala, 33rd Proceedings of Electronic Components and Conference, (1983), pp. 189-192.

[46] A. Bischoff and F. Aldinger, 34th Proceedings of Electronic Components and Conference, (1984), pp. 411-417.

[47] J. Kurtz, D. Cousens, and M. Dufour, 34th Proceedings of Electronic Components and Conference, (1984), pp. 1-6.

[48] M. L. Minges, Electronic Materials Handbook Volume 1 PACKAGING (ASM International, 1989).

Chapter 4

Resistor materials and high K dielectric materials

4.1 Introduction

LTCCs incorporate a variety of materials, and allow the integration of different functions. They enable components and substrates to be made smaller as well as multifunctional, and offer benefits when compared with other technologies (for example, printed circuit board technology) [1]. The representative dissimilar materials combined in LTCCs are the resistor materials used for termination in order to prevent reflected noise, and the dielectric materials with high dielectric constants used in the bypass capacitor layer for the power supply. The purpose of the bypass capacitor is to maintain a constant supply voltage by supplying a charge to the IC when its supply voltage falls momentarily due to simultaneous switching noise. Here a capacitor with a high speed of response is desirable, and in order to reduce wiring inductance that slows down speed of response, it is necessary to locate it in proximity to the IC. Therefore, for achieving low inductance, it is very effective to embed it in the substrate [2].

This chapter covers general information about the resistors introduced into LTCCs, as well as issues concerning ruthenium oxide – glass composites which are typical resistor materials for LTCCs. In addition, the development of dielectric materials for LTCCs that have high dielectric constants is a very active field at present, with a variety of materials being suggested and with the development of new materials underway. At the moment it is difficult to summarize all technologies in development. For this reason, we will focus on the development issues involved when incorporating material with a high dielectric constants in LTCCs that use copper as a conductor.

The technology development for the processes involved in incorporating high frequency functions besides capacitors, such as filters, and the materials development required for fulfilling the various functions, is covered in the outlook for the future in Chapter 10.

4.2 Resistor materials

Resistors can be broadly classified into two kinds; those formed on the surface of the LTCC, and those formed inside. When forming the resistor on the surface, one method is to polish the surface of the LTCC after firing, and to form a resistor film using a thin film technique such as sputtering, vapor deposition or the like. Another method is to print a thick film resistor paste on the surface of the as-fired substrate and then fire it. Internal resistors formed within the substrate are obtained as follows. A conductor pattern is formed on the surface of the ceramic green sheet using screen printing then resistive paste is printed in specific positions. Then after alignment of these printed green sheets, they are superimposed and made into a single green sheet laminate with heat and pressure, and are then fired at a high temperature of around 900°C.

When forming a thin resistor film on the surface, the costs are high since it is necessary to polish the surface of the substrate. However, because the shape and dimensions of the resistor can be controlled easier, it is easy to achieve the designed resistor values. In addition, with resistors that are formed on the surface of the LTCC, the value of resistance can be adjusted with laser trimming. When resistors are formed within layers, the substrate contracts when firing, so not only is it difficult to control the dimensions of the resistor, there is also a concern that the value of resistance may change due to reactions with the ceramic during firing.

$$\text{TCR (ppm/°C)} = \Delta R/(R_{25} \cdot \Delta T) \times 10^6$$

R_{25}: Value of resistance at 25°C

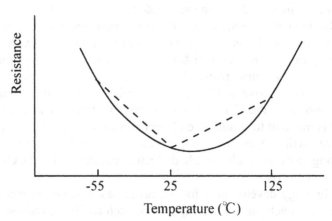

Figure 4-1   Conceptual diagram of the TCR of a resistor.

The characteristics of resistors that are of interest are their value of resistance and the temperature dependency of their value of resistance (TCR - Temperature Coefficient of Resistance). As shown in Table 4-1, TCR is a temperature gradient for the value of resistance, and with room temperature as the boundary, two values, for the high temperature side (25 to 125°C) and the low temperature side (25 to -55°C), are provided. The required resistance value differs depending on the application, but for the TCR, a value of around ±100 to ±300 ppm/°C is required, irrespective of the value of resistance.

Thick film resistor paste consists of inorganic components that form the resistor (conductive material and low softening point glass [PbO-SiO_2-B_2O_3-Bi_2O_3 glass]) and an organic vehicle (binder, plasticizer, solvent or the like). The conductive materials used differ in terms of the intended value of resistance and firing process atmosphere as shown in Table 4-1. Table 4-2 shows typical materials used as thin film resistors [3, 4].

Table 4-1   Inorganic conductive materials in thick film resistor paste and their resistance specifications.

Firing atmosphere	Sheet Resistance ($\Omega/\square$)	TCR (ppm/°C)	Conductive material [Resistivity $\Omega \cdot cm$ (25°C)]
Air	1 - 1 M	±250 to ±300	Ag/Pd (PdO) [4×10^{-5}]
	10 - 10 M		RuO_2, IrO_2 [4×10^{-5}]
	10 - 1 M	±50 to ±300	$Pb_2Ru_2O_6$,$SrRuO_3$, $Bi_2Ru_2O_7$ [2.3×10^{-2}]
Nitrogen	10 - 10 K	±100	LaB_6 [17×10^{-6}]
	5 - 1 M	±250	SnO_2-Sb

Table 4-2   Thin film resistor materials and their resistance specifications.

Material	Sheet Resistance value ($\Omega/\square$)	TCR (ppm/°C)
Ni-Cr	50 to 500	< ±20 to ±150
Cr_2O_3	100 to 1,000	±25 to ±100
Cr-SiO_2	100 to 1,000	< ±20 to ±200
TaN	100 to 1,000	-60 to -150
W-Ru	100 to 1,000	+120 to +300
Ti	100 to 1,000	-100 to +100

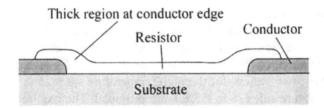

Figure 4-2   Cross section of a thick film resistor formed on a substrate (after firing).

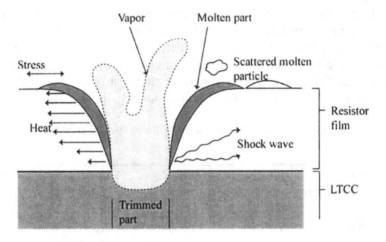

Figure 4-3   Conceptual diagram of laser trimming.

There are three main methods used to control the value of resistance. These are 1) controlling the dimensions of the resistor, 2) controlling material composition, and 3) trimming. By changing the dimensions of the resistor film formed on the substrate (width and height), it is possible to change the value of its resistance. Depending on the circumstances, changing the film thickness can be effective, but in order to achieve a reliable and stable resistance value, it is desirable to have a specific film thickness for the resistor. As shown in Figure 4-2, when planning the resistance value based on the shape, it is necessary to pay consideration to the electrode edge-effect since the film thickness of the resistor changes with proximity to the electrode [5]. It is also necessary to control the resistance of the interface through the reaction between the resistor and electrode interface. Also, since the resistive paste is a composite of conductive material and glass material as noted above, the value of resistance can be controlled easily by changing the composition ratio of both, and by changing the kind of conductive material used. The two methods above for controlling the value of resistance are performed before forming the resistor film. In contrast, the third method, trimming, is a technique for changing the value of resistance by trimming the resistor film with a laser after it has been formed. Figure 4-3 is a conceptual diagram of the laser trimming method. When using the laser trimming method with a precisely programmed laser, it is easy to achieve a stable resistance value within 0.5% in a short time. In order to achieve a value of resistance in accordance with the settings, it is necessary consider the damage to the resistor caused by the laser.

4.2.1 Ruthenium oxide/glass material

This section describes the characteristics and issues associated with the material ruthenium oxide/glass that is widely used in thick film resistor paste.

Ruthenium oxide is a conductive oxide that displays metallic behavior. In the atmosphere it does not change up to 1,025°C, and above 1,400°C it breaks down and partially vaporizes. It is formed if ruthenium or ruthenium chloride are heated in an oxygen airflow, and it is generally available on a commercial basis as a powder with a particle size of 10 nm or more. Table 4-3 shows the representative characteristics of ruthenium oxide.

Table 4-3 The representative characteristics of ruthenium oxide.

Density	7.06 g/cm^3
Mole density	18.85 cm^3
Coefficient of thermal expansion (20 to 300°C)	6.32×10^{-6}/°C
Resistivity (25°C)	4×10^{-5} Ω • cm
TCR	5×10^3 ppm/°C

For the glass material in ruthenium oxide and the glass composite, it is common to use lead borosilicate glass with a low softening point with the aim of making a resistor with an optimal firing temperature of around 900°C. In insulating materials of this type where conductive material particles are dispersed in the material matrix, in order to achieve contact between the conductive particles, it is necessary for the volume fraction of the conductive material to be greater than 10 to 30 vol% if the particle diameter of each of the materials is roughly the same. However, if the particle diameter of the conductor is much smaller than the particles of the insulating material as with ruthenium oxide/glass (refer to Figure 4-4), it is reported that the minimum volume fraction is 3 to 10 vol% in order to form a contact path for the conductive particles [6, 7, 8]. With material resistance R, an equivalent circuit model can be expressed with the following equation.

$$R = R_c + R_{gb} + R_g$$

R_c: resistance of the conductive material, R_{gb}: resistance of the grain boundary, R_g: resistance of the glass

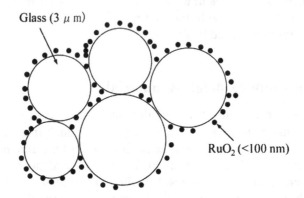

Figure 4-4   Schema of the powder mixing state before firing.

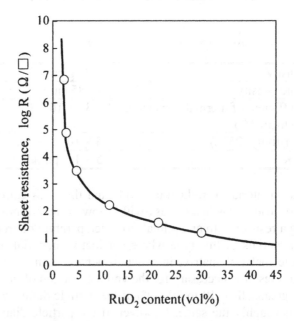

Figure 4-5   The composition dependence of the electrical resistance of ruthenium oxide/glass resistors.

From the macro point of view, it is possible to control electrical resistance by changing the volume ratio of the ruthenium oxide and glass, as shown in Figure 4-5 [9]. Various models have been suggested for the conduction mechanism present between the ruthenium oxide particles in the glass layer, and the matter is currently being debated [10, 11, 12, 13, 14].

4.2.2 The thermal stability of ruthenium oxide

Ruthenium oxide is thermodynamically stable in the atmosphere up to around 1,400°C, but at a higher temperature or at low oxygen partial pressure, metallic ruthenium is the stabilized phase (refer to Figure 4-6) [15, 16]. Similarly to ruthenium oxide, metallic ruthenium is conductive, but as the electrical resistance (ruthenium oxide: 4×10^{-5} $\Omega \cdot$ cm, ruthenium 7.2×10^{-6} $\Omega \cdot$ cm) and molar volume (ruthenium oxide: 19 cm^3, ruthenium 8.2 cm^3) of the substances differ, if ruthenium oxide transforms into metallic ruthenium in the resistive paste during firing, it causes an extreme change in the resistance value and volumetric shrinkage, while generating stress inside the resistor. For this reason, it is necessary to understand fully the oxidation-reduction reaction of ruthenium oxide (Ru-RuO$_2$) [17].

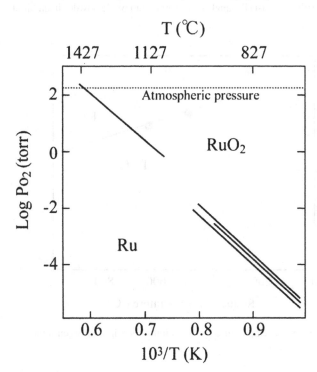

Figure 4-6   Ru-RuO$_2$ phase diagram (temperature, oxygen partial pressure dependency) (Ref. [15]).

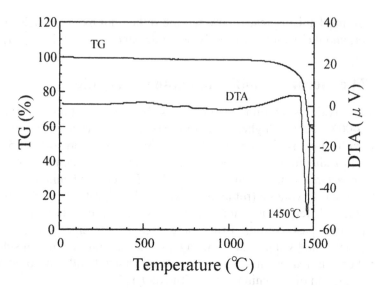

Figure 4-7   The results of TG-DTA analysis of ruthenium oxide powder in an air atmosphere.

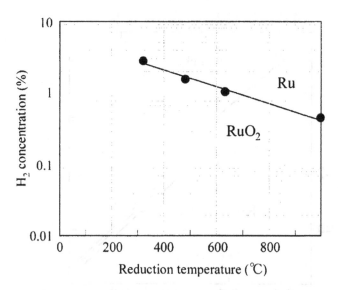

Figure 4-8   The reduction temperature of ruthenium oxide in a nitrogen atmosphere mixed with hydrogen.

The reduction temperature of ruthenium oxide can easily be found using differential thermal analysis TG-DTA. Figure 4-7 shows the results of TG-DTA analysis of ruthenium oxide powder with purity greater than 99.9%, and average particle diameter 0.05 μm in an air atmosphere. The DTA curve shows a large endothermic peak, and the temperature with weight reduced by 24% on the TG curve is the observed reduction temperature (around 1,450°C). This result closely matches Figure 4-6. In addition, the reduction temperature falls markedly in an atmosphere with hydrogen added as in Figure 4-8 [18, 19, 20].

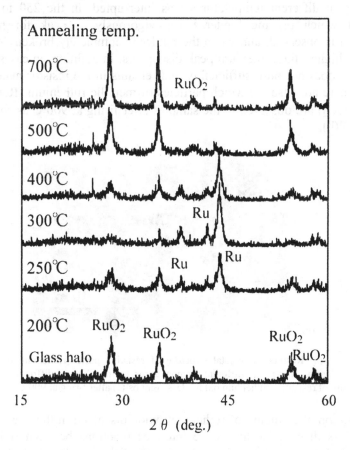

Annealing temp. (°C)	Peak detected
− 200°C	RuO$_2$, Glass halo
250 − 400°C	Ru, RuO$_2$, Glass
500°C −	RuO$_2$, Glass

(a)

Since the firing of resistive paste (ruthenium oxide/glass) is usually carried out in the atmosphere without the introduction of reducing gas, this issue may be considered irrelevant to deoxidation. However, when the organic binder that is added to resistive paste combusts at high temperature, it creates a reducing atmosphere surrounding the ruthenium oxide. Figure 4-9(a) shows the results of X-ray diffraction of samples of resistive paste when the firing at different temperatures was interrupted. In the 250 to 400°C range at which organic binder burns vigorously, a peak for metallic ruthenium is observed, and when the binder is completely broken down, at 500°C or higher, the ruthenium peak disappears. In addition, in cases where the binder does not burn sufficiently and remains in the resistor material as carbon, there are cases in which particles of metallic ruthenium (RuO_2 + C → Ru + CO_2) are observed in the samples after firing at 900°C as shown in Figure 4-9(b).

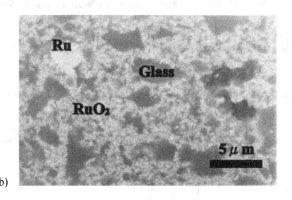

(b)

Figure 4-9   (a) Results for crystal structure of resistive paste (ruthenium oxide/glass composite) during firing at different temperatures using X-ray diffraction. (b)The microstructure of resistors (ruthenium oxide/glass composite) after firing at 900°C.

In addition, the amount of ruthenium oxide dissolved in the glass is very little at less than 0.5%, and since interface reactions between ruthenium oxide and glass are not observed, the impact of the interfacial phenomenon on the value of resistance is small [21, 22].

4.3   High K dielectric material

To introduce high K dielectric material into an LTCC, two main methods are known. One is to make the dielectric material into a paste, and to form it using screen printing on the surface of the insulating material green sheet, and the other is to make the dielectric material itself into a green sheet and to laminate it with green sheets of insulating material to form a layer within a multilayer structure. With either method, the first requirement of the material is that it can be fired at a low temperature of around 900°C since it will be cofired with the insulating material and conductive material. Besides this, a high dielectric constant and high insulation are required. Among promising materials under development, the three groups of materials shown in Table 4-4 are known as materials likely to meet these requirements [23, 24, 25, 26, 27].

Table 4-4   Dielectric materials introduced in LTCCs.

Material type	Representative example	Dielectric constant
Glass/Ceramic composite	• $BaTiO_3$ - glass • $CaZrO_3$ - $SrTiO_3$ - glass	1,000 25
Liquid-phase sintered ceramics	• $BaTiO_3$ - LiF, • BaSrTiO3 -Li_2CO_3, B_2O_3 etc.	3,000 to 4,000 2,000 to 3,000
Pb system relaxor	• Pb $(Zn_{1/3}Nb_{2/3})O_3$ • Pb $(Fe_{1/3}W_{2/3})O_3$ • Pb $(Mg_{1/3}Nb_{2/3})O_3$ etc.	22,000 10,000 15,000

The dielectric materials for low temperature firing shown in Table 4-4 react easily with the glass in the insulating material, and their dielectric constant falls markedly. In order to bring out their dielectric characteristics, it is necessary to make improvements from the point of view of materials and processes, for example by providing a barrier layer between the insulating material and dielectric material.

4.3.1   Issues with low oxygen partial pressure atmosphere firing (point defects and semiconductor formation)

Besides their impact on mechanical properties, lattice defects in the crystal also have a significant effect on the electrical characteristics of the material. Oxide crystals with large forbidden band widths also show semiconducting properties due to the impact of these lattice defects. The electronic behavior of lattice defects is explained below. As shown in Figure 4-10 (a), when defects are formed by atoms moving from their normal lattice position to the interstice, the same quantity of lattice vacancies and interstitial atoms are formed. This sort of defect morphology is called the Frenkel disorder. Furthermore, defects such as shown in Figure 4-10 (b), in which lattice vacancies of cations and anions are formed at the same time are known as the Schottky disorder. In addition to these defects, when impurities present in the ceramic are substituted in the normal lattice position making substitutional solid solutions, or dissolve in the interstice to form interstitial solid solutions, their atomic value is controlled to maintain electrical neutrality and lattice defects are formed.

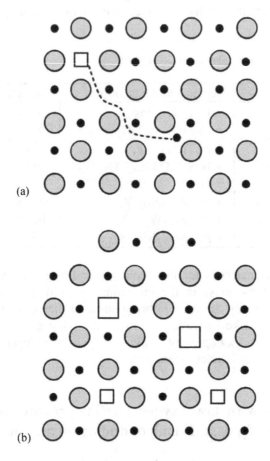

(a)

(b)

Figure 4-10   Conceptual diagram of lattice defects (a) Frenkel disorder, (b) Schottky disorder.

The following Kröger-Vink notation is widely used for point defects. The region A at the bottom right of Figure 4-11 shows an element at a location where a defect is formed. If it is formed in the interstitial, it is written as I. In region B the element that exists at A site after the defect is introduced is shown. If there is a vacancy, V is shown. The region C at top right shows the valance balance. If the valence balance is +, • is used, and if it is -, ' is used; valence is indicated by the number of each symbol [28].

ex.

$$B_A^c \quad V_o^{\cdot\cdot} \quad Ti_{Al}^{\cdot} \quad Ca_I^{\cdot\cdot} \quad Mg_{Al}'$$

Figure 4-11   Kröger-Vink notation.

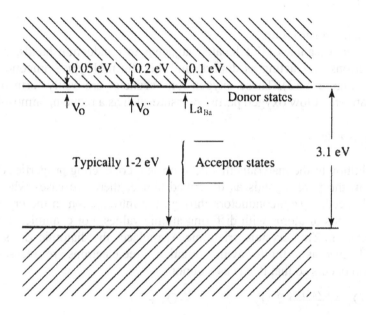

Figure 4-12   Levels of donors and acceptors for defects of $BaTiO_3$.

The following section covers the electrical characteristics caused by lattice defects
(1) Positive ion vacancies
Positive ion vacancies (V_m) are located adjacent to valence bands, and by trapping electrons from the valence band they become neutral, causing holes in the valence band (refer to Figure 4-12). For this reason, positive ion vacancies act as acceptors, becoming P-type semiconductors. FeO is a representative example, and since these crystals are stable because Fe is stoichiometrically lacking, Fe vacancies (V''_{Fe}) are formed in the crystal.

$$nil \rightarrow V''_{Fe} + 2h^{\cdot}$$

Other known P-type oxide semiconductors formed due to positive ion vacancies are NiO and Cu_2O.

(2) Positive ion interstitial atoms

Positive ions (M_I) that invade and are present between the lattice positions ionize from a neutral state, and release electrons to the conduction band (refer to Figure 4-12). For this reason, positive ion interstitial atoms act as donors, and the crystals form N-type semiconductors. Representative examples are ZnO and SnO_2.

$$nil \rightarrow Zn^{\cdot}_I + e'$$

(3) Oxygen ion vacancies

Oxygen ion vacancies (V_O) located adjacent to conduction bands, release free electrons thus acting as donors, and they form N-type semiconductors (refer to Figure 4-12). These oxygen ion vacancies are easily formed with heat treatment at low oxygen partial pressure such as a reducing atmosphere.

$$nil \rightarrow V^{\cdot\cdot}_O + 2e'$$

In addition to the materials that develop semiconducting properties due to defects in the pure crystals as described above, there are cases where the materials become semiconductors through the introduction in the crystal of different kinds of atoms with differing atomic values. For example, if Al_2O_3 is dissolved in ZnO crystals, triatomic Al replaces the diatomic Zn, so that $Al^{\cdot}_{Zn}$ is formed as in the following equation. Free electrons are released and ionization occurs so that it acts as a donor.

$$Al_2O_3 \xrightarrow{(2ZnO)} 2Al^{\cdot}_{Zn} + 2O_O + \frac{1}{2}O_2 + 2e'$$

In addition, if Li^+ ions are dissolved in NiO, Li'_{Ni} is formed which works as an acceptor, making a P-type semiconductor.

$$Li_2O + \frac{1}{2}O_2 \xrightarrow{(2NiO)} 2Li'_{Ni} + 2O_O + 2h^{\cdot}$$

As described above, oxides can easily become semiconductors through the effects of the atmosphere of heat treatment, and the introduction of different kinds of element. The following is an example of formation of a semiconductor using the most common dielectric ceramic, barium titanate. Figure 4-13 shows the experimental value for the conductivity of barium titanate when the heat treatment temperature and oxygen partial pressure of the ambient atmosphere is changed [29, 30]. As shown in the Figure, conductivity shows a concave curve. The low oxygen partial pressure side (the left side) of the lowest point of the curve in the Figure shows an N-type semiconductor, while the high oxygen partial pressure side (the right side) shows a P-type semiconductor. Figure 4-14 shows the variation in conductivity and defect density when a donor and acceptor are added to BaTiO$_3$. As in the Figure, by adding donors and acceptors, it is possible to change the stability region (the position of the lowest point of the conductivity curve) of each type of semiconductor [31]. Incidentally, as shown in the following equation, lanthanum oxide (La$_2$O$_3$) acts as a donor, while manganese oxide (MnO) acts as an acceptor with regard to BaTiO$_3$.

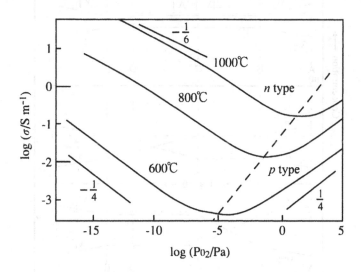

Figure 4-13   The relationship between heat treatment temperature and oxygen partial pressure in the atmosphere, and conductivity of undoped BaTiO$_3$.

$$La_2O_3 \xrightarrow{\;(2BaTiO_3)\;} 2La^{\bullet}_{Ba} + 2O_O + \frac{1}{2}O_2 + 2e'$$

$$MnO + \frac{1}{2}O_2 \xrightarrow{\;(BaTiO_3)\;} Mn''_{Ti} + 2O_O + 2h^{\bullet\bullet}$$

When copper is used as a conductive material in LTCCs, since firing at low oxygen partial pressure is required, besides the dielectric material becoming semiconductive, and the insulating properties of the material being lost, there is also a risk that an increase in dielectric loss will occur. While paying due consideration to the oxygen partial pressure of the firing environment, it is necessary to increase resistance to reducibility by introducing additives that work as acceptors, shifting the transition point between N-type and P-type semiconductors to the low oxygen partial pressure side.

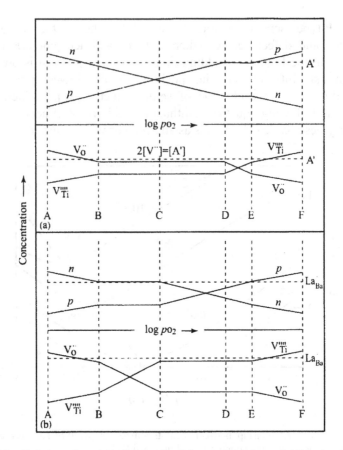

Figure 4-14   Relative values (calculated values) for conductivity and defect density when a donor and acceptor are added to BaTiO$_3$.

References

[1] Y. Imanaka, Material Technology of LTCC for High Frequency Application, Materials Integration, Vol. 15, No. 12, (2002), pp. 44-48.

[2] Y. Imanaka, T. Shioga, and J. D. Baniecki, "Decoupling Capacitor with Low Inductance for High-Frequency Digital Applications,"*FUJITSU Sci. Tech. J.*, Vol. 38, No. 1 June, (2002), pp. 22-30.

[3] P. J. Holmes, R. G. Loasby, Handbook of Thick film Technology, Electrochemical Publication Limited, (1976), pp. 147-150.

[4] R. W. Vest, "Materials Science of Thick Film Technology," Ceramic Bull., Vol. 65, No. 4, (1986), pp. 631-636.

[5] Electric Materials Handbook Volume 1 Packaging ASM International, Materials Park, OH, (1989), pp. 344.

[6] T. Inokuma, Y. Taketa and M. Haradome, "The Microstructure of RuO_2 Thick Film Resistors and the Influence of Glass Particle Size on Their Electrical Properties," IEEE Transactions on Components, Hybrids, and Manufacturing Technology, Vol. CHMT-7, No. 2, June, (1984), pp. 166-175.

[7] P. F. Carcia, A. Ferretti and A. Suna, "Particle size effects in thick film resistors," J. Appl. Phys. Vol. 53, No. 7, July, (1982), pp. 5282-5288.

[8] J. Lee and R. W. Vest, "Firing Studies with a Model Thick Film Resistor System," IEEE Transactions on Components, Hybrids, and Manufacturing Technology, Vol. CHMT-6, No. 4, Dec., (1983), 430-435.

[9] A. Kusy, "Chains of Conducting Particles that Determine the Resistivity of Thick Resistive Films," Thin Solid Films, Vol. 43, (1977), pp. 243-250.

[10] A. Kusy, "On the Structure and conduction Mechanism of Thick Resistive Films," Thin Solid Films, Vol. 37, (1976), pp. 281-302.

[11] G. E. Pike and C. H. Seager, "Electrical properties and conduction mechanisms of Ru-based thick-film (cermet) resistors,"J. Appl. Phys., Vol. 48, (1977), pp. 5152-68.

[12] D. P. H. Smith and J. C. Anderson, "Electrical Conduction in Thick Film Paste Resistors," Thin Solid Films, Vol. 71, (1980), pp. 79-89.

[13] P. J. S. Ewen and J. M. Robertson, "A percolation model of conduction in segregated systems of metallic and insulating materials: application to thick film resistors," J. Phys. D: Appl. Phys., Vol. 14, (1981), pp. 2253-68.

[14] D. S. McLachlan, M. Blaszkiewicz and R. E. Newnham, "Electrical Resistivity of Composites," J. Am. Ceram. Soc., Vol. 73, No. 8, (1990), pp. 2187-2203.

[15] V. K. Tagirov, D. M. Chizhikov, E. K. Kazenas and L. K. Shubochkin, Zh. Neorg. Khim., Vol. 20, No. 8, 2035 (1975); Russ. J. Inorg. Chem. (Engl. Transl.) Vol. 20, No. 8, (1975), pp. 1133.

[16] W. E. Bell and M. Tagami, "High-Temperature Chemistry of the Ruthenium-Oxygen System ," Physical Review, Vol. 67, Nov. (1963), pp. 2432-2436.

[17] J. W. Pierce, D. W. Kuty, and J. R. Larry, "The Chemistry and Stability of Ruthenium-Based Resistors," Solid State Technology, Oct. (1982), pp. 85-93.

[18] M. Hiratani, Y. Matsui, K. Imagawa and S. Kimura, "Hydrogen Reduction Properties of RuO_2 Electrodes," Jpn. J. Appl. Phys. Vol. 38, (1999), pp. L1275-L1277.

[19] L. K. Elbaum and M. Wittmer, "Conducting Transition Metal Oxides: Possibilities for RuO_2 in VLSI Metallization, " J. Electrochem. Soc.: Solid-State Science and Technology, Oct. , (1988), pp. 2610-2614.

[20] Y. Kaga, Y. Abe, M. Kawamura and K. Sasaki, "Thermal Stability of RuO_2 Thin Films and Effects of Annealing Ambient on Their Reduction Process," Jpn. J. Appl. Phys. Vol. 38, (1999), pp. 3689-3692.

[21] A. Prabhu, G. L. Fuller and R. W. Vest, "Solubility of RuO_2 in a Pb Borosilicate Glass," J. Am. Ceram. Soc., Vol. 57, No. 9, (1974), pp. 408-409.

[22] Y. M. Chiang, L. A. Silverman, R. H. French and R. M. Cannon, "Thin Glass Film between Ultrafine Conductor Particles in Thick-Film Resistors," J. Am. Ceram. Soc., Vol. 77, No. 5, (1994), pp. 1143-1152.

[23] J. V. Biggers, G. L. Marshall and D. W. Strickler, "Thick-Film Glass-Ceramic Capacitors," SOLID STATE TECHNOLOGY, May, (1970), pp. 63-66.

[24] H. Mandai et al., Ceramic Science & Technology Congress Proceedings, (1990), pp. 391.

[25] B. E. Walker Jr., R. W. Rice, R. C. Pohanka and J. R. Spann, "Densification and Strength of BaTiO3 with LiF and MgO Additives, " Ceramic Bull., Vol. 55, No. 3, (1976), pp. 274-276.

[26] J. M. Haussonne, G. Desgardin, PH. Bajolet, and B. Raveau, "Barium Titanate Perovskite Sintered with Lithium Floride, " J. Am. Ceram. Soc., Vol. 66, No. 11, (1983), pp. 801-807.

[27] S-J. Jang, W. A. Schulze and J. V. Biggers, "Low-Firing Capacitor Dielectrics in the System $Pb(Fe_{2/3}W_{1/3})O_3$-$Pb(Fe_{1/2}Nb_{1/2})O_3$-$Pb_5Ge_3O_{11}$," Ceramic Bull., Vol. 62, No. 2, (1983), pp. 216-218.

[28] P. Kofstad, Nonstoichiometry, Diffusion, and Electrical Conductivity in Binary Metal Oxides, Wiley, New York (1972).

[29] N. H. Chan and D. M. Smyth, "Defect Chemistry of $BaTiO_3$,"J. Electrochem. Soc. , Vol. 123, No. 10, (1976), pp. 1584-85.

[30] N. H. Chan R. K. Sharma and D. M. Smyth, "Nonstoichiometry in Undoped $BaTiO_3$,"J. Am. Ceram Soc., Vol. 64, No. 9, (1981), pp. 556-62.

[31] D. M. Smyth, Prog. Solid State Chem., Vol. 15, (1984), pp. 145-71.

Part 2

Process technology

Chapter 5

Powder preparation and mixing

From Chapter 2 through Chapter 4 we focused on material technology of LTCCs. In the following Chapters 5 through 8 we will consider the process technologies used for producing LTCCs. However, since there is some overlap with the general manufacturing processes of ceramics, we will look exclusively at processes unique to LTCCs, and at the phenomena that appear to be peculiar to each process. The overall manufacturing process for LTCCs is shown in Figure 1-3 in Chapter 1. They are manufactured in accordance with the procedures of powder preparation, mixing, casting, blanking, via forming, pattern forming, laminating, and firing.

The details of each process are covered in the following chapters.

5.1 Introduction

In the manufacture of LTCCs, the first process is powder preparation. Powder preparation involves the selection of raw materials, and after various pretreatments are performed on each material (thermal, chemical, and mechanical), the materials are combined to achieve the intended material. When mixing the powder for LTCCs, inorganic ceramic materials (ceramic powder and glass powder) and a great variety of organic materials are used. In general, the organic materials used are binders, plasticizers, dispersing agents, and solvents. However of the raw materials used, the organic materials are those where careful selection has the most significant impact on formability.

After the various materials are weighed, the raw material is mixed in a ball mill or the like. The process for making the slurry used for casting is a mixing process; it is the same as the manufacturing process for common ceramics. The following section covers the various special features of the inorganic materials and the organic materials that must be understood when manufacturing LTCCs. The section also covers the characteristics of the slurry.

5.2 Inorganic ceramic materials

As we explained in Chapter 1, the main types of LTCC dielectric ceramics are glass/ceramic composites and crystallized glass. The characteristics required of the raw materials differ for each type.

With glass/ceramic composites, since at least two kinds of inorganic material powder – ceramic and glass powder – are used, it is desirable for the difference in charging characteristics from the other materials to be minimal. If a charge is formed at the surface of the two kinds of powder with differing positivity and negativity, attraction results, and the different types of material powder aggregate. If the two kinds of powder form the same charge with a significant difference in the quantity of electric charge, aggregation of the same kind of powder can be expected through repulsion. When selecting the particle diameter and specific surface area of the powder, it is necessary to pay consideration to the amount of binder added for shaping, and the fluidity of the glass when firing. If the particle diameter of the powder is small and the specific surface area is large, it is necessary to increase the amount of binder for shaping with the result that the filling ratio of the powder in the compact is low. Thus when firing, there is a tendency for the firing shrinkage to increase. If the firing shrinkage rate is significant, in general there can be a lot of variation depending on the location of shrinkage, so for this reason it is desirable for the particle diameter of the powder to be large, and the specific surface area to be small. Furthermore, since it is important for the liquidation of the glass to occur efficiently, glass powder with small particles, where liquidation occurs readily, is appropriate, while ceramic powder with a comparatively large particle diameter is beneficial as the glass penetrates the capillaries between the ceramic particles and fluidizes readily. Incidentally, while it makes it more difficult to achieve a sintered body with high density, it is effective to use fine ceramic particles in order for example to increase the strength of the glass/ceramic composite, and it is necessary to determine the most appropriate particle diameter in accordance with the purpose of the product. In order to achieve a homogeneous compact and to enable uniform firing, it is desirable for both powders to have a narrow particle size distribution and for the size of the particles to be similar. If there is a broad particle size distribution, parts of the composition will be irregular and uneven voids will be formed.

In crystallized glass type LTCCs, since one kind of ceramic powder is used, it is not necessary to consider its charging characteristics. However, in order to achieve a homogenous structure, it is desirable to have a sharp distribution of particle size, and it is necessary to decide the powder particle size and specific surface area taking into account formability and sinterability in the same way as with glass/ceramic composites.

5.3 Organic materials

The commonly known organic materials used for the manufacture of LTCCs are binders for maintaining the strength of the compact and increasing its formability; plasticizers that give the slurry its rheological properties and that give the compact plasticity and flexibility; dispersing agents, the additives that provide control of the pH of the slurry and the charge of the particle surface, provide steric hindrance between particles, and disperse agglomerated particles into primary particles; antifoaming agents to prevent the occurrence of foam in the slurry; surface treatment coupling agents to improve poor wetability of ceramic powder by lowering its surface tension; and nonaqueous organic solvents. Basically, these organic materials are substances that are added to all the intermediate products – slurry, green sheet, and green laminate body – and they are important additives on which the characteristics of the intermediate products depend. However, since they must be completely eliminated in the final firing process, basically it is desirable that a minimal amount of each additive is used. Table 5-1 shows the representative organic materials used as additives [1, 2]. The details of the various additives are covered in the following sections.

Table 5-1   Representative organic materials used as additives in LTCCs.
Nonaqueous

Solvent	Binder	
Acetone	Cellulose acetate butyrate	
Benzene	Butyrate	
Bromochloromethane	Cellulose nitrate	
Diacetone	Petroleum resin	
Butanol	Polyethylene	
Ethanol	Polyacrylic ester	
Propanol	Polymethyl methacrylate	
Methyl isobutyl ketone	Polyvinyl alcohol	
Toluene	Polyvinyl butyral	
Trichloroethylene	Vinyl chloride	
Xylene	Polymethacrylate	
	Ethyl cellulose	
	Abietic acid resin	
Plasticizer	Dispersing agent	Wetting agent
Butyl benzil phthalate	Fatty acid (glyceryl trioleate)	Alkylaryl polyether alcohol
Dibutyl phthalate	Fish oil	Ethyl ether of polyethylene
Butyl stearate	Synthetic surfactants	glycol
Dimethyl phthalate	(benzenesulfonic acid)	Ether ethylphenyl glycol
Methyl abietate	Oil-soluble sulfonates	Polyoxyethylene ester
Compound of phthalic ester	Oleic acid ethylene oxide adduct	Monoolein acid glycerine
Derivative of polyethylene glycol	Sorbitan trioleate	Triolein acid glycerine
Tricresyl phosphate	Phosphate ester	
	Steric acid amide ethylene oxide adduct	
	Menhaden fish oil	
	Natural sardine oil	
	Octadiene	

Aqueous

Solvent	Binder	
Water (Isopropyl alcohol can also be used) Wax, silicone and nonionic surface active agent are used as antifoaming agents.	Acrylic polymer Emulsion of acrylic polymer Ethylene oxide polymer Hydroxyl ethyl Cellulose Methyl cellulose Polyvinyl alcohol Isocyanate Wax wetting agent Aqueous urethane Salt of methacrylic acid copolymer Wax emulsion Emulsion of ethylene-vinyl acetate copolymer	
Plasticizer	Dispersing agent	Wetting agent
Butyl benzil phthalate Dibutyl phthalate Ethyltoluene sulfamido glicentin Polyalkyl glycol Triethylene glycol Tri-n-butyl phosphate Polyol	Phosphate Phosphoric acid complex salt Natural sodium Aryl sulfonic acid Acrylic oligomer	Nonionic octyl phenoxyethanol Nonionic surface active agent

5.3.1 Binder

Various functions and roles are required of the binders used in LTCCs in the several processes of producing the slurry, green sheet, and laminated body. In the mixing process for producing the slurry, the solubility of the binder with the other organic materials and wetability with the inorganic material powder are important factors, so that the type and kind of binder has an effect on the viscous behavior of the slurry. Ensuring that defects such as cracks and the like do not occur in the green sheet is important in the forming process in which the slurry is made into sheets. In preventing flaws, the drying conditions when casting the sheets, as well as the mechanical properties of the binder material itself, are important factors. After green sheet fabrication and until the lamination process, it is important that the green sheet has good mechanical strength with excellent handleability and flexibility – when stripping the carrier film used in forming the green sheet, during blanking of the sheet, and when making the via holes. In addition, when storing the green sheet, dimensional changes in the sheet due to changes in the external environment (temperature, humidity, and so on) should be minimal. In the printing process when the electrode paste is printed in the circuit pattern, the binder must not dissolve or become wet due to the solvent in the paste. Since the binder plays the role of bonding the green sheets together in the heat pressing step of the lamination process, it must have excellent adhesive properties. Furthermore, the binder must not gasify during thermocompression bonding leaving air spaces between layers of green sheet, and it is important that pores appropriate to the green sheet are formed. In the firing process, the binder gradually undergoes thermal decomposition, and finally all residues must be completely eliminated.

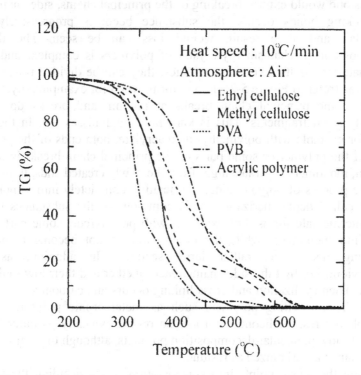

Figure 5-1 The thermal decomposition behavior of representative binder resins (Air atmospheric heating).

Figure 5-1 shows the thermal decomposition behavior of representative binder resins when heated in the air atmosphere. As the figure shows, the final thermal decomposition end temperature differs according to the type of binder, and acrylic polymer has the best pyrolytic properties. When copper is used as a wiring material in LTCCs, the thermal decomposition behavior of the binder must be borne in mind particularly since a low oxygen concentration atmosphere is used and even after firing, binder tends to remain as residue in the ceramic [3].

In general, polymeric materials break down when heated at high temperature, and some or all of the ingredients disappear. When heated in the atmosphere, first some of the polymers oxidize due to the oxygen present in the atmosphere, and oxides of the hydroperoxide group, carbonyl group and so on are formed. In this way, the polymers are broken and their molecular weight falls. Next, when the temperature nears the boiling point of the oxides generated, they evaporate and their weight is seen to decrease. When heated in an inert atmosphere in which oxygen is not present, oxidation does not occur. However when the temperature rises to a certain level, as one would expect, breaking of the principal chains, side chains, and cross-linking bonds occurs, the substance becomes progressively low-molecular, and as a result, weight loss can be seen. The thermal decomposition mechanism (pyrolysis) of polymers is complex, and while each material has its own characteristics, they can be divided roughly into two forms (refer to Figure 5-2). In one form, when for example polyethylene is heated, the principal chain breaks at random and breaks down into variously sized fragments. This is known as random session. In the other form, for example with polymethyl methacrylate, both ends of the principal chain of the polymer or some parts of the principal chain break. In a chain reaction, monomers separate singly from the newly created ends, through the opposite process of polymerizing, dissociating completely into a monomer. This is called depolymerization. It is common for the substances to take these intermediate forms. For example, with polystyrene, some parts break down into monomers, while the other parts do not become monomers, remaining instead as rather large sections. In addition, as with polyacrylonitrile, first the side chains react together and their low molecules are lost. Then cyclization and cross linking occurs and carbonized residue is formed. Thermosetting polymers such as metamorphic polyester, epoxy resin, phenol resin, silicone resin and so on are typically partially broken down into low molecular decomposition products, although the major part of the substance is carbonized as residue.

Taking the above points into consideration, when deciding the type of binder to use, it is necessary to determine the degree of polymerization and molecular weight besides the sort of polymer.

5.3.2 Plasticizer

Plasticity is the quality of being deformed permanently by a small force. When plasticizer is added to various kinds of binder, the glass transition point and melting point of the resin falls, giving it flexibility and making it easy to shape. The following qualities are required of the plasticizers for green sheets used in LTCCs.

(1) Excellent compatibility with binder resin, (2) high boiling point and low vapor pressure, (3) high plasticizing efficiency, (4) stability with regard to heat, light, and chemical substances, (6) excellent flexibility at low temperature, (7) the plasticizer does not migrate readily when in contact with other materials, and so on.

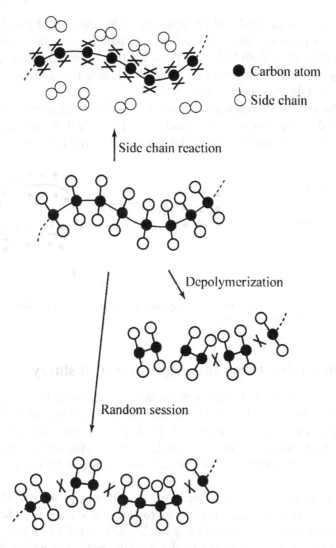

Figure 5-2   The mechanism of thermal decomposition (pyrolysis) of polymers.

There are two techniques for plasticization, external plasticization and internal plasticization. With both techniques, the distance between each polymer increases, and the bond strength between the polymers weakens making micro-brownian motion occur readily, resulting in flexibility (refer to Figure 5-3). Internal plasticization is a method that uses chemical means such as copolymerization, substitution of atomic groups and so on. For example, as vinyl chloride contains chlorine atoms, its intermolecular force is very strong, but when it is copolymerized with a substance with relatively long side chains such as vinyl acetate, its glass transition point falls. External plasticization is a method that provides plasticity through physical mixing. By mixing a primary plasticizer that is compatible with the resin with a secondary plasticizer with little compatibility, thus combining plasticizer and the polar group of the polymer chain and enveloping the polymer chain, the space between the polymers is increased.

In shaping LTCC green sheets, dibutyl phthalate is frequently used since it has good compatibility with binder, and satisfactory results can be obtained when it is added to the binder in proportions of 10 to 30%.

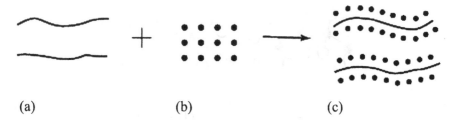

(a)   (b)   (c)

Figure 5-3   Diagram of plasticization; (a) Polymers, (b) plasticizer, (c) plasticized polymers (space between molecules increases).

5.3.3   Dispersing agent and dispersibility of slurry

There are two factors in dispersing the ceramic particles in the slurry reliably [4]. One uses the repulsion of the charges of the particles, while the other uses the effect of steric hindrance of the dispersing agent absorbed in the particle surface. When the former is used, controlling the surface potential of the particles is an important point in determining the stability of the slurry [5].

Electrical neutrality is established near the surfaces of the particles and a charge with the opposite sign, equivalent to the surface charge, gathers like a cloud in the form of ions around the particle surface (refer to Figure 5-4). In this structure, the layer attached to the particle surface is called the Stern layer, and outside that, the layer that is present as a result of equilibrium distribution through the balance between electrostatic attraction and diffusion force from thermal motion, is called the diffuse electric double layer. If an external electric field is applied to this kind of dispersion system, the particle and diffuse electric double layer are drawn electrically to an electrode of the opposite sign, and relative motion occurs at the boundary of a certain "slide plane". The potential of this slide plane is called the zeta potential, and it is used as the scale of surface potential [6, 7, 8].

The zeta potential is determined by the charge density of the particle surface, and by the concentration and type of dispersing agent and so on. The surface potential of particles differs significantly with the acidity or basicity of the material, and even within the same material, it varies slightly with the method of manufacture, the heat treatment conditions and so on (refer to Table 5-2 and Figure 5-5). Although the surface potential varies according to the pH of the solution, the pH at which the apparent surface potential is zero is called the isoelectric point, and aggregation of the particles occurs at this pH [9, 10, 11].

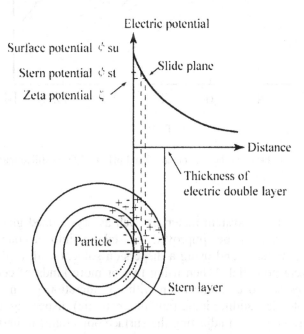

Figure 5-4   Ion distribution around a particle in a solution.

Table 5-2   The isoelectric points of various kinds of oxides and hydroxides.

MgO	12.4	CeO_2	6.75
La_2O_3	10.4	TiO_2	6.7
ZrO_2	10-11	SnO_2	6.6
BeO	10.2	SiO_2	1.8
CuO	9.4	WO_3	-0.5
ZnO	9.3	$Co(OH)_2$	11.4
α-Al_2O_3	9.1	$Ni(OH)_2$	11.1
α-Fe_2O_3	9.04	$Zn(OH)_2$	7.8
Y_2O_3	9.0	γ-$FeO(OH)$	7.4
γ-Al_2O_3	7.4-8.6	α-$FeO(OH)$	6.7
Cr_2O_3	7.0	$Al(OH)_3$	5.0-5.2

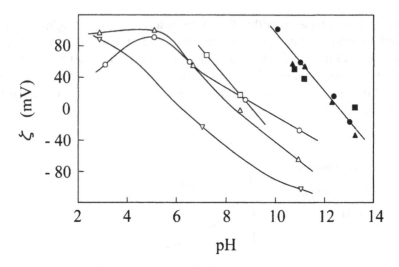

Figure 5-5 The relationship between the zeta potential and pH of Al_2O_3 (unfilled mark) and MgO (filled mark).

For LTCCs, with single system materials such as crystallized glass and the like, dispersibility can be improved by taking into account the equipotential of the material and using a dispersion solvent with a pH that results in a high zeta potential. When using two or more kinds of ceramic particles, it is necessary to combine them with consideration given to the charged electric potential, adding ionic (anionic, cationic) dispersing agents according to requirements, and adjusting the surface potential. Furthermore, the method of adding a nonionic dispersing agent to form a polymer adsorption layer (refer to Figure 5-6), and achieving dispersion through the effect of steric hindrance, is effective irrespective of the surface potential of the particles [12, 13, 14].

In general, as shown in Figure 5-7 dispersing agents are compounds that in one molecule share a hydrophilic group part and a lipophilic group part. Dissolved in a solvent, they adhere to the particle surface, orient themselves, and by reducing the surface tension they display their various characteristics. The HLB value (Hydrophile Lipophile Balance) value is an index that shows the balance of strength between the hydrophilic group and the lipophilic group [15, 16].

HLB value = $\{M_w/(M_w+M_o)\} \times 20$
  M_w: Hydrophilic molecular weight
  M_o: Lipophilic molecular weight

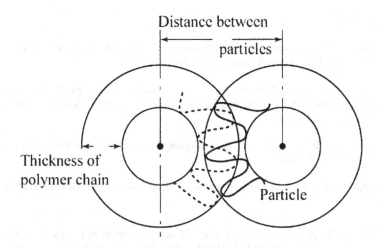

Figure 5-6   Schema of the steric stability effect of polymer adsorption.

With organic slurries, the HLB value has a conspicuous impact on the density of the compact. For example, with alumina powder with a particle size of 1 μm or less, a dispersing agent with an HLB value with a low lipophilic molecular weight prevents aggregation of the powder and improves its dispersibility. If the amount of dispersing agent added is increased, the solution properties change abruptly above a certain concentration. If a small amount is added, it remains in a state of monomolecular dissolution in the solvent, but above a certain concentration it is probable that colloid aggregates and micelles are suddenly formed. This minimum concentration at which micelles are formed is called the critical micelle concentration (CMC). Dispersing agents must be used at a level above CMC. If the dispersing agent is used at a lower concentration, the lowering effect of surface energy and other characteristics worsen abruptly [17]. However, if the amount added is excessive, the micelles become entangled with each other and the surface activation effect is lost.

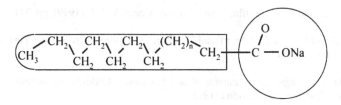

Figure 5-7   The basic structure of dispersing agent (Left: lipophilic part, Right: hydrophilic part) [Example: Fatty acid sodium].

References

[1] R. E. Mistler, "Tape Casting; The Basic Process for Meeting the Needs of the Electronic Industry," Ceramic Bull., Vol. 69, No.6, (1990), pp. 1022-1026.

[2] K. Saito, "Use of Organic Materials for Ceramic-Modeling Process-Binder, Deffloculant, Plasticizer, Lubricant, Solvent, Protective Colloid-," J. of the Adhesion Society of Jpn, Vol. 17, No. 3, (1981), pp. 104-113.

[3] K. Saito, "Organic Materials for Ceramic Molding Process," Bull. Ceram. Soc. Jpn., Vol. 18, No. 2, (1983), pp. 98-102.

[4] G. D. Parfitt, Dispersion of Powders in Liquids, American Elsevier, New York, (1969), pp. 315.

[5] R. E. Mistler, R. B. Runk, and D. J. Shanefield, eramic Fabrication Processing Before Firing, Edited by G. Y. Onoda and L. L. Hench, Wiley, New York, (1978), pp. 411-48.

[6] A. Kitahara, "Electrokinetic Potential-Zeta Potential," Bull. Ceram. Soc. Jpn., Vol. 19, No. 1, (1984) pp. 38-42.

[7] K. Tamaribuchi, and M. L. Smith, J. Colloid Interface Sci., Vol. 22, (1966), pp. 404.

[8] L. Gouy, J. Phys., Vol. 9, 457 (1910); Ann. Phys., Vol. 7, (1917), pp. 129; D. L. Chapman, Phil. Mag., Vol. 25, (1913), pp. 475.

[9] G. A. Parks, Chem. Rev., 65, (1965), pp. 177.

[10] R. H. Ottewill and A. Watanabe, Kolloid-Z., Vol. 107, (1960), pp. 132.

[11] J. B. Kayes, J. Colloid Interface Sci., Vol. 56, (1976), pp. 426.

[12] E. W. Ficher, Kolloid-Z., Vol. 160, (1958), pp. 120.

[13] R. H. Ottewill, T. Walker, Kolloid-Z. u. Z. Polymere, Vol. 227, (1968), pp. 108.

[14] D. H. Napper and A. Netschey, "Steric Stabillization of Colloidal Particles," J. Colloid Interface Sci., Vol. 37, No. 3, (1971), pp. 528-535.

[15] R. W. Behrens and W. G. Griffin, J. Soc. Cosmet. Chem. Vol. 1, (1949), pp. 311

[16] J. T. Davies and E. K. Rideal, Interfacial Phenomena, Academic Press, New York and London, (1963), pp. 343-447

[17] K. Shinoda, T. Nakagawa, B. Tamamushi, and T. Isemura, Colloidal Surfactants, Academic Press, New York and London (1963)

Chapter 6

Casting

6.1 Introduction

After mixing various kinds of ceramic powder and organic materials such as binder, plasticizer, dispersing agent, and solvent according to a specific composition using a ball mill, the resulting slurry is formed into green sheets. The green sheet is an intermediate product before creating the final product, but the quality of the LTCC after firing depends to a large extent on the quality of the green sheet. This is because the shaping of the multilayer substrate such as hole punching, via opening, circuit wiring, multilayer formation and so on is performed before firing in the green sheet state. In order to achieve the intended characteristics of the green sheet, it is necessary to pay careful consideration to the quality and casting conditions of the slurry used for casting.

This section provides an outline of green sheet casting and offers a number of points to consider when preparing the slurry. In addition, it gives details considering the characteristics of the green sheet in the steps from the forming process leading up to the printing process. The previous chapter covered the issues to bear in mind when preparing the raw materials and mixing them to obtain the slurry. This section also covers the characteristics of slurry that must be considered when casting it, after mixing it with the ball mill.

6.2 Casting equipment

The equipment used for casting green sheets for use in LTCCs is shown in Figure 6-1. Currently, a variety of casting equipment is being manufactured but in general, the equipment consists of a carrier film conveyor, casting head, slurry dispenser, drying area, and sheet take-up unit [1, 2]. The carrier film conveyor fulfills the role of conveying the plastic carrier film, fed from a roll, to the casting head. Since the plastic film is the carrier of the cast sheet, it is desirable that it has no wrinkles, and travels in a straight line at an even speed. At the casting head, the ceramic slurry is dispensed onto the carrier film. The slurry dispenser is for volumetric feed of the slurry to the casting head in order to produce the ceramic green sheet reliably and continuously [3, 4]. The drying area drives off the solvent in the cast ceramic slurry to produce a dried sheet [5, 6]. Drying normally uses infrared heaters or hot air. The drying temperature profile is adjusted taking into account the drying rate of the slurry and the speed of the carrier film. The sheet take-up unit picks up the dried ceramic green sheet in a roll. Some take-up units remove the green sheet from the carrier film while others take up the carrier film as well. PET (polyethylene terephthalate) film is commonly used for carrier film, and according to requirements, a silicone release agent is applied in order to improve peelability.

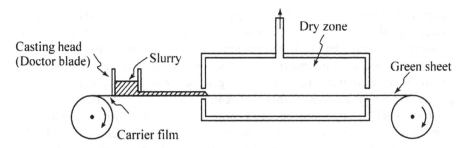

Figure 6-1   Conceptual diagram of green sheet casting equipment.

Figure 6-2 is a conceptual diagram of the casting head which is the core of casting machines. This is the technique known as the doctor blade method, in which a gap of a certain size is formed with a doctor blade. Slurry is fed with a quantitative supply to a slurry tank, and with the movement of the carrier film, the ceramic slurry is extruded from the head forming a sheet. The slurry flows naturally from the gap formed by the doctor blade due to the pressure determined by the height of the liquid level, and is extruded due to the movement of the film [7, 8]. A relationship between these factors in accordance with the following equation has been suggested. In order to control the thickness of the sheet, it is necessary to control the blade gap, carrier film speed, the height of the fluid level in the slurry tank and so on [9, 10]. When the slurry is extruded, it is also necessary to take into consideration the action of force pushing the blade up making the blade gap slightly larger.

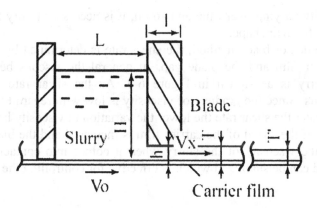

Figure 6-2   A typical casting head and a casting head for stabilizing sheet thickness. Manufacturing parameters (Vx: Slurry flow speed, T: Slurry thickness immediately after passing the blade, T' Green sheet thickness after drying).

$$t = \alpha(\frac{\rho g H}{12\mu l V_0} h^3 + \frac{h}{2})$$

t: sheet thickness, g: gravity, l: blade thickness, α: drying shrinkage rate, H: slurry fluid level height,
V_0: carrier film speed, ρ: density, μ: slurry viscosity
h: doctor blade gap

Controlling external factors such as those covered above, as well as controlling the internal material factors of the slurry itself to achieve appropriate viscous behavior, is important for stabilizing sheet thickness and sheet quality.

6.3 Slurry characteristics

The slurry used for doctor blade casting is a non-Newtonian fluid containing a high concentration of ceramic powder and resin that shows thixotropy, as with the conductive paste used for printing explained in the next chapter. Thixotropy is the quality where, when a significant shear is applied, viscosity falls and when left to stand viscosity increases. This is an important characteristic for casting (see Figure 6-3). In other words, when the slurry comes into contact with the casting head its viscosity falls while its fluidity increases. After casting, viscosity increases preventing unnecessary distortion of the sheets. However, since thixotropy shows unstable behavior with viscosity varying over time and so on, it is necessary to dry the sheets quickly and fix their shape.

With the doctor blade method, the shear rate is determined by the speed of the carrier film and the blade gap. In general the viscous behavior of ceramic slurry is as shown in Figure 6-3. As the shear rate increases, viscosity falls. Since the gradient of viscosity is low with regard to the shear rate, the greater the shear rate the lower the variation in viscosity becomes. It follows that if the speed of the carrier film is increased and the blade gap is decreased, the viscosity of the slurry when it comes into contact with the casting head can be stabilized which is effective for controlling the thickness of the sheet.

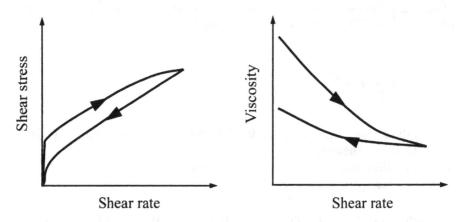

Figure 6-3 The flow curve and viscosity curve of ceramic slurry.

In general, since the slurry experiences more shear stress with a greater shear rate as shown in Figure 6-3, it is known that if the powder contained in the slurry is granular or acicular in shape, it becomes oriented [11, 12]. Figure 6-4 plots the difference in shrinkage rates in the x and y axes in the thickness of green sheets, after casting and after each green sheet is fired. There is a tendency for the difference in firing shrinkage rates in the x and y axes to be greater with thinner sheets. It is likely that this is due to the particles in the sheet becoming oriented more readily in thinner sheets. In addition, casting with a high shear rate is effective in the elimination and homogeneous dispersion of the air phase that can be thought to be inherent in the slurry, while it is also effective in achieving a homogeneous pore structure in the finished green sheet.

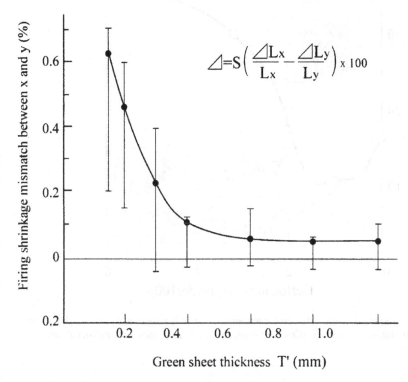

$$\varDelta = S \left(\frac{\varDelta L_x}{L_x} - \frac{\varDelta L_y}{L_y} \right) \times 100$$

Figure 6-4 Chart of the difference in shrinkage rates in the x and y axes in the thickness of green sheets, after casting and after each green sheet is fired.

It is desirable for the ceramic powder in the slurry to be dispersed homogeneously in the solvent, and it is necessary to optimize the amount of dispersing agent added as explained in the previous chapter. As shown in Figure 6-5, when only a small amount of dispersing agent is added, air phase remains in the lumps of powder forming agglomerates. Conversely, when too much is added, it flocculates. Either case leads to the viscosity of the slurry increasing. It is necessary to discover the amount of dispersing agent required to lower the viscosity and distribute the powder through experiments.

Although the viscosity of the slurry itself cannot be varied significantly, by leaving the prepared slurry to stand under reduced pressure, the air phase in the slurry is eliminated while the solvent constituents are driven off, making it possible to control the viscosity.

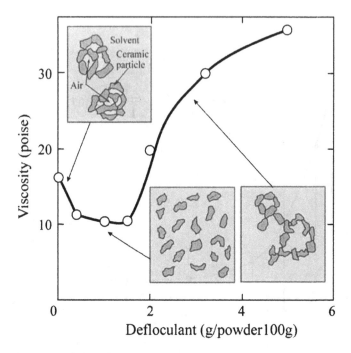

Figure 6-5 The relationship between the amount of dispersing agent added to the slurry and the conformation of the powder (Conformation: From left, agglomeration, deflocculation, flocculation).

6.4   Green sheet

Various characteristics are required of green sheets cast using the doctor blade method in order to achieve stable characteristics in the final fired product. The following sections describe the requirements for green sheets, methods of evaluating the green sheet, and the various materials, manufacturing methods, and environmental parameters that have an effect on the characteristics of green sheets. They also cover processes that are effective in achieving stable characteristics.

6.4.1   Characteristics required of green sheets

Green sheets consist of ceramic powder and binder, resin materials such as plasticizer and dispersing agent and voids. It is desirable if the various ingredients are dispersed evenly and the structure is three dimensionally homogeneous in the x and y axes and z axis. From the point of view of macro-structure, the thickness of the green sheet should be uniform (since the thickness of the green sheet is stipulated as the thickness of one layer of the insulating layer in a multilayer substrate, this is an important factor in the consistency of the characteristic impedance of transmission circuits). The sheet should show little anisotropy, and there should be no macro defects such as scratches or cracks. The surface should have excellent smoothness, and it is desirable if there is no difference in the microstructure of the top and bottom of the sheet.

In each manufacturing process, the following issues must be taken into account in order to achieve intermediate products with the intended characteristics. When leaving the green sheets to stand and when storing them, it is desirable for dimensional changes and changes in characteristics to be kept to the minimum. If the dimensions change after the circuit pattern is printed on the surface of the green sheet or after blanking or punching, the green sheet will not match those above and below it in the multilayer structure, resulting in conductive defects and the like. In addition, the green sheet must have the mechanical strength (hardness and elongation) to withstand the handling involved in each process. If the green sheet is removed from the carrier film, the sheet may stretch or even break in some cases. As explained above, the cutting and punching processes may cause open or short circuits due to the shape of the holes formed. For this reason, sheets with good workability are desirable, in which scraps do not stick to the sheets and their shape during processing can be kept close to that of the design. In the printing process the surface of the green sheet should show minimal roughness. Also the conductive paste formed by being pressed through the printing screen should have excellent adherence with the green sheet. After printing, the vehicle contained in the paste is required to have good compatibility with the green sheet, penetrating and being absorbed by it without the printed pattern becoming blurred. For lamination, the sheets must have elasticity and thermoplasticity, but while the layers should adhere together well, it is desirable that they do not distort any more than is necessary.

6.4.2 Green sheet evaluation methods

At present, the methods of evaluation for green sheet compacts are by and large based on onsite experience and no absolute evaluation methods have been established. The methods shown below are those generally used in practice, and they can be considered to be effective for quality control of green sheets.

(1) Surface roughness [13]

Since the surface roughness of green sheets is influenced by the size and aggregation of the ceramic powder particles, the dispersibility of the raw materials in the green sheet can be considered to be one relevant scale. However, the surface condition of the sheets is influenced by the processing conditions such as the viscometric property and drying conditions of the coating and the like. This influence differs with the thickness of the sheets, and the thinner the coating, the faster the film dries. Therefore, since leveling of the film does not proceed sufficiently, the surface roughness increases. Increasing the amount of resinous constituents in the raw materials makes the surface smoother without affecting dispersibility. For this reason, it is fruitless to compare absolute values and to discuss dispersibility. However, when changing one material or manufacturing process parameter, making a relative comparison of the state of dispersion from measured values can be effective. To increase printability when forming fine wiring on the green sheet, it is also important to evaluate sheet surface roughness.

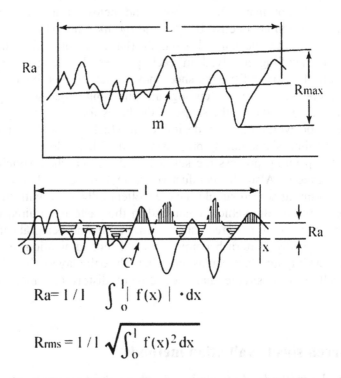

$$Ra = 1/1 \int_{0}^{1} | f(x) | \cdot dx$$

$$Rrms = 1/1 \sqrt{\int_{0}^{1} f(x)^2 \, dx}$$

Figure 6-6 The notation for surface roughness (R_a: center line average roughness, l: measured height, C: center line, R_{rms}: root mean squared roughness, R_{max}: maximum height).

Surface roughness is found through numeric conversion of a surface contour profile measured with a surface roughness gauge at a fixed scanning distance. There are various methods to determine the numeric value, but generally the three methods shown in Figure 6-6 are used. R_a is the center line surface roughness, achieved by finding the integral of the surface contour profile and dividing it by the scanning distance. R_{rms} shows the root mean squared. The highest point in the profile is shown as maximum roughness R_{max}.

The degree of reflection of the material surface correlates with surface roughness as shown in Figure 6-7, and as a testing method it is effective for product control of green sheets [14]. Normally the index known as glossiness is used. Light from a standard light source is shone on the surface of the sample, and its glossiness is the relative numeric value for the reflected light picked up by a photodetector. If the incident light is fixed and the angle of the photodetector is changed, the reflection coefficient is at its maximum when the photodetector is rotated 60 to 75 degrees. At this angle, sensitivity to the surface roughness of the sample is at its highest point, and since the reflection coefficient changes significantly with even a small surface roughness in the sample, it is possible to convert subtle differences in surface roughness into numeric values.

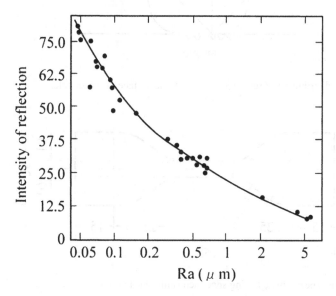

Figure 6-7 The relationship between intensity of reflection and surface roughness R_a.

(2) Tensile testing [15]

Tensile strength and elongation is generally evaluated using tensile testing to show the mechanical properties of the green sheets. The specimen configuration shown in Figure 6-9 used for tensile testing of plastic materials is commonly used. A low crosshead speed of around 0.5 mm/min is used for tensile testing. Figure 6-10 shows the rate of strain dependency of the stress – strain line when the green sheet undergoes tensile testing. Tensile strength and elongation varies significantly at the time of rupture according to the rate of strain. This suggests the need for very careful handling of the green sheets during processing. Tensile property test specimens are punched out of green sheet with a mold and are formed into a prescribed shape. However if there are any notches or the like in the testing region, it is unlikely that accurate measurements will be taken so that considerable care must taken when making samples.

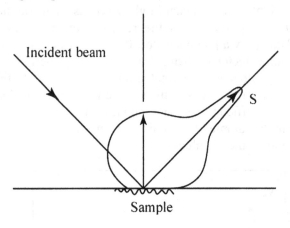

Figure 6-8 The distribution of reflected light from the surface of the sample.

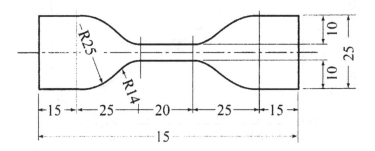

Figure 6-9 Green sheet tensile testing specimen (unit: mm).

(3) Gas permeability

This method is generally used for evaluating the airtightness of plastic film. An airtight container in which outside air is in contact with a film plane is filled with gas (helium) at a certain pressure. The gas leaks from the film plane and the time taken to return to the pressure outside (atmospheric pressure) is measured [16, 17]. Then the gas permeability K is calculated using the D'Arcy equation. When this method is applied to green sheet, there is no particular significance attached to the measured value for gas permeability itself. However, due to the fact that a numeric value representing the amount of voids and the structure of the green sheet can be calculated makes this an effective evaluation method for comparing the quality of green sheets.

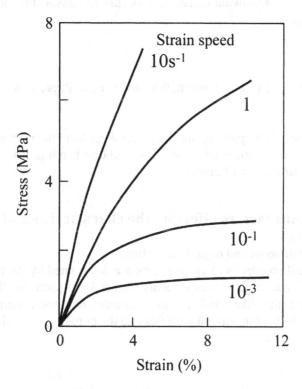

Figure 6-10 Green sheet tensile testing results (rate of strain dependency)

$$K = V \cdot h \cdot \mu/(S \cdot \Delta P \cdot \tau)$$

V: quantity of flow through the sample (m^3)
h: thickness of the sample (m)
μ: viscosity of the gas $(N \cdot s/m^2)$
S: area of the sample (m^2)
ΔP: pressure difference (N/m^2)
τ: time taken to return to normal pressure (s)

Besides the methods noted above, study of the microstructure of the surface using SEM, and techniques for studying the fracture surface using the frozen section method (the method whereby soft green sheet containing plastic is chilled at low temperature and is cut to achieve a brittle fracture surface) are frequently used. Evaluation using mercury porosimetry where mercury is injected into the sample is an effective technique for comparing the void characteristics between samples [18, 19]. Since the diameter, distribution and amount of voids are measured directly through the relationship in the following equation, it is easy to convert the measurement results into figures.

d = 2σcosθ/P

d: void diameter, σ: surface tension, θ: contact angle, P: pressure

As noted below, by picking out the green sheet from the relevant location and measuring the amount of weight loss through heating, it is possible to examine the distribution of binder.

6.4.3 Various factors affecting the characteristics of green sheets

(1) The effect of voids and organic constituents
The mechanical properties of green sheets are determined by the mechanical properties of the organic constituents themselves such as the binder, plasticizer and the like, and by the geometrical arrangement (particle dispersibility) of the constituent particles and the porosity of the sheets.

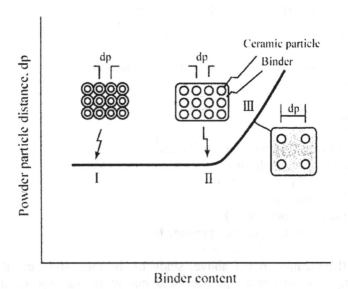

Figure 6-11   The relationship between the amount of binder in the green sheet and the distance between ceramic particles [Ref. 19].

Figure 6-11 is a conceptual diagram of the relationship between the amount of binder in the green sheet and the distance between ceramic particles [20]. Even if the amount of binder is increased, the distance between the ceramic particles does not change until a specific amount of binder is added (range I). In this range, the binder is largely positioned in the gaps between the ceramic particles arranged in the green sheet, and as the amount of binder added increases, it fills in the gaps. Then the surface of the particles is coated with a thin binder layer. II in the figure shows the critical point where the added binder has filled in all the gaps between the particles. In range III after the critical point, a binder phase is formed between the particles as more binder is added. As a result, the distance between the ceramic particles increases. From range I to II, there is a tendency for the density of the green sheet to increase with the amount of binder added. On the other hand, in range III the density of the green sheet falls instead when more binder is added. The qualities of the green sheet approach those of the resin itself, and its thermal and mechanical properties such as elongation, and thermoplasticity and dimensional stability at high temperature worsen markedly. A structure typical of range I to II is appropriate for green sheets used in LTCCs.

The bonding between particles can be considered to be an effect of the organic constituents. As shown in Figure 6-12 (a), since the mechanical properties of green sheets vary significantly according to the degree of polymerization of the binder resin, quality control of the resin is important, for example, by finding its average molecular weight before use and so on. The plasticizer significantly changes the rheological behavior of the binder resin, and as described in the previous chapter, since it changes the polymer structure by increasing the spaces between polymers, it has a great impact on the mechanical properties of the resin. As shown in Figure 6-12 (b), as the amount of plasticizer is increased, the green sheet generally becomes softer, and a tendency for its tensile strength to fall and elongation to increase can be seen. Figure 6-13 shows the relationship between the amount of organic material added (binder + plasticizer) and the various characteristics of a green sheet with a 75/25 ratio of resin/plasticizer. With increasing amounts of organic material, tensile strength and elongation increase. Additionally, surface roughness and gas permeability fall with the increase in the amount of organic material. This is caused by the gaps between particles in the green sheet being filled in by the resinous constituents.

By optimizing the amount of dispersing agent, the dispersibility of the ceramic particles in the slurry improves so that the structural homogeneity of the green sheet increases [21]. Figure 6-14 shows an example of the glossiness of the green sheet increasing due to the addition of an optimal amount of dispersing agent [22].

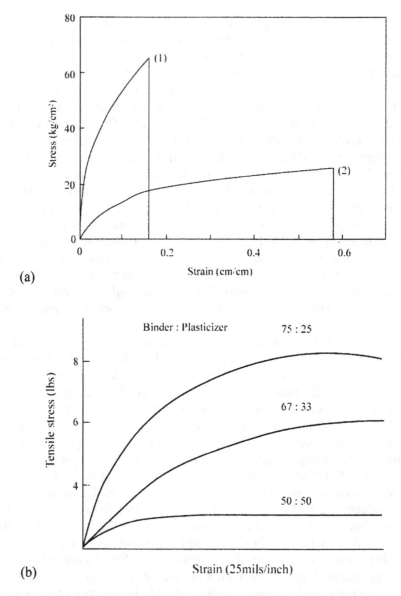

Figure 6-12   The mechanical properties of green sheets (a) impact of the degree of polymerization of the binder resin [the degree of polymerization is higher for (1)], (b) impact of binder/plasticizer ratio.

The diameter of the ceramic particles has an effect on the strength of the green sheet, and in general, the smaller the particle size, the greater its strength. However, with small particles the surface area of the particles increases so that more binder must be added. Since the structure begins to approach range III in Figure 6-11 this is not desirable for the green sheet.

To achieve dimensional stability in green sheets, they frequently undergo pressure treatment after casting. In this case the distance between particles is decreased due to the increased pressure, and the strength of the green sheet increases.

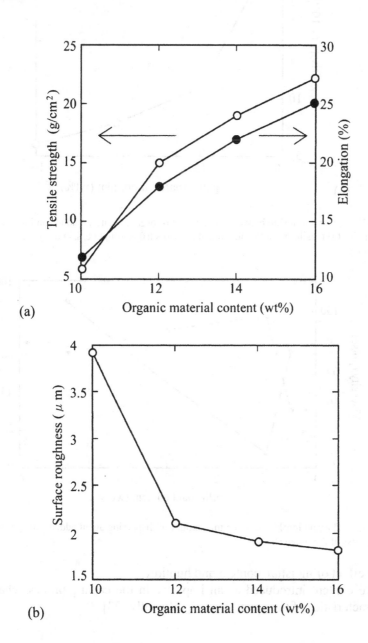

(a)

(b)

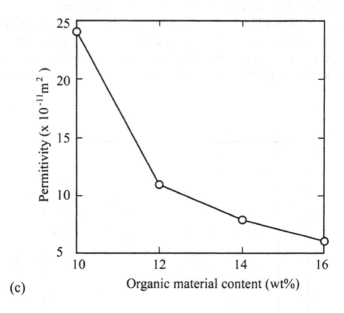

(c)

Figure 6-13   The relationship between the amount of organic material added and green sheet characteristics (a) tensile strength and elongation, (b) surface roughness, and (c) gas permeability.

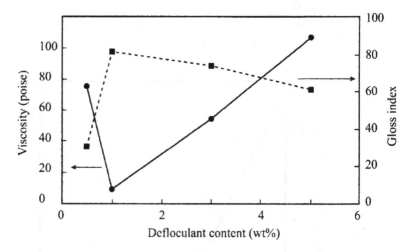

Figure 6-14   The relationship between the amount of dispersing agent added, slurry viscosity, and glossiness of the green sheet.

(2) The effect of moisture content and humidity

Moisture content introduced as an impurity in the casting process changes the characteristics of the green sheet significantly [23].

Figure 6-15 shows the relationship between the amount of contamination by moisture in the slurry formed by the composition of the solvent and the slurry viscosity, and the various characteristics of a green sheet. When resin which does not dissolve in water is used as binder or plasticizer, part of the resin is precipitated by the moisture contaminant, significantly changing the characteristics of the dispersion medium, degrading the dispersibility of the ceramic particles in the slurry, and raising the viscosity of the slurry. As a result, the structure of the green sheet becomes heterogeneous, and voids of uneven size occur readily. At the same time, the binder resin has a tendency to coagulate locally so that its density and tensile strength falls while gas permeability, surface roughness, and elongation increase.

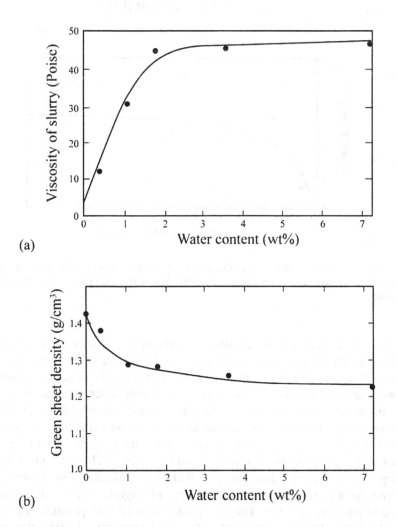

(a)

(b)

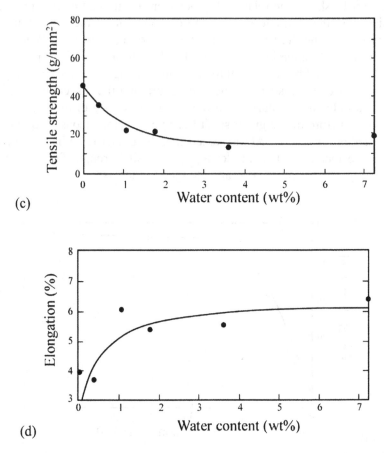

Figure 6-15   The relationship between the amount of moisture content in the slurry and its various characteristics (a) slurry viscosity, (b) green sheet density, (c) tensile strength, and (d) elongation.

Moisture content introduced into the slurry as an impurity consists mainly of moisture forming an adsorption layer on the surface of the base powder, so that it is important to exercise care regarding its storage conditions (temperature and humidity), and to pretreat it with water removal processes and so on before mixing the powder. Figure 6-16 shows the relationship between storage humidity and the amount of water adsorption of the powder surface in the case of alumina and borosilicate glass (Pyrex glass), commonly used as a constituent material of LTCCs. The amount of water adsorption of alumina is low, however that of borosilicate glass is more than 10 times greater than alumina, and it depends significantly on the storage humidity and storage time. The water saturation of both powders increases with the humidity. In the example of storage humidity of 60%, the water absorption of the alumina powder reaches saturation equilibrium in around 24 hours, while the borosilicate glass powder requires 72 hours to reach equilibrium. Since the amount of water saturation through adsorption is affected by the degree of basicity of the materials themselves as well as by the particle size of the powder, specific surface area and so on, it is important to grasp the amount of water saturation through adsorption of each different powder.

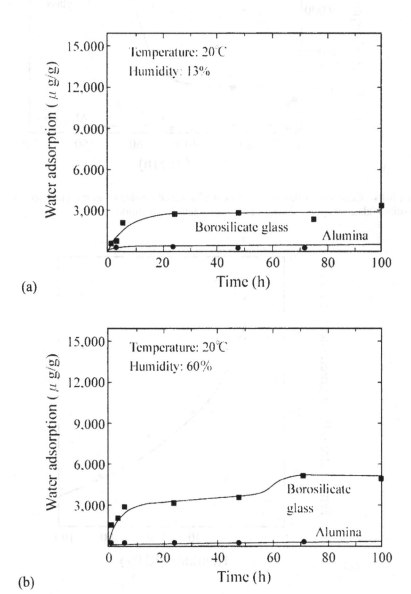

(a)

(b)

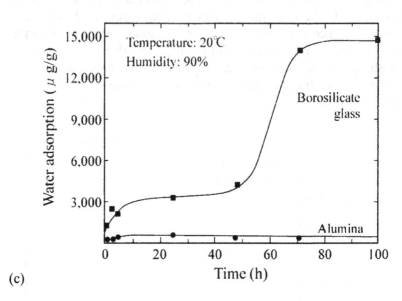

(c)

Figure 6-16   Change over time of the amount of water adsorption of alumina powder and borosilicate glass powder under various storage humidity conditions.

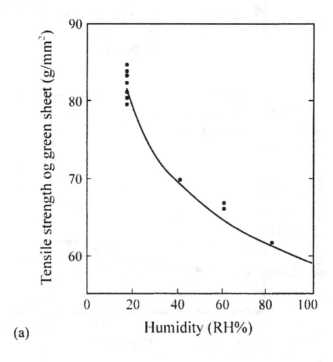

(a)

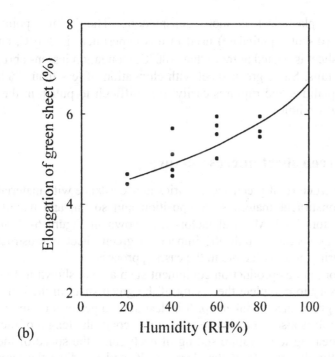

(b)

Figure 6-17   The relationship between green sheet storage humidity and green sheet characteristics; (a) tensile strength, and (b) elongation.

There is also a close relationship between green sheet storage environment humidity and the mechanical properties of the green sheet after casting as shown in Figure 6-17. The tensile strength and elongation of green sheets that show satisfactory mechanical properties immediately after casting also changes in accordance with the storage humidity. For example, in the hole punching process, there are cases where via hole process defects occur due to the effect of elongation when punching is performed with the same processing conditions. Therefore it is necessary to store the materials at a specific storage humidity.

(3) The effect of temperature
In order to eliminate the solvent left in the green sheet after casting and the moisture content described above, green sheets are frequently dried in a thermostatic oven after casting. However, besides the solvent and moisture content, the green sheet constituents also volatilize at the same time during this process and since this significantly changes the mechanical properties of the green sheet, it is necessary to set the temperature of the heat treatment carefully. Figure 6-18 shows the relationship between the heat treatment temperature and weight loss of green sheet constituents, and green sheet tensile strength and elongation. Although weight loss of both binder and plasticizer can be seen with an increase in temperature of the heat treatment,

the loss of plasticizer is more conspicuous. The boiling point of the plasticizer (di-butyl phthalate) used in this experiment is 340°C, and when the green sheet is heated at more than 100°C, elongation lessens abruptly and flexibility falls. Since green sheet with elongation of less than 1% has poor handling qualities and ruptures easily, it is difficult to punch in the process for opening via holes.

6.4.4 Green sheet microstructure

The microstructure of green sheets varies in accordance with material factors such as constituent materials, composition and so on, and manufacturing process factors [24]. Material factors are shown in Figure 6-11 and other figures. This section details the impact on green sheet microstructures of manufacturing process factors in the casting process.

With continuous production equipment such as that shown in Figure 6-1, it is necessary to complete the drying of the green sheet in the heating zone before the green sheet is taken up. It is desirable to perform drying as slowly as possible. It is also necessary to take into account the length of the heating zone, the heating temperature setting of each zone, the speed of the carrier film, and the drying rate of the slurry itself, and to adjust the temperature profile accordingly.

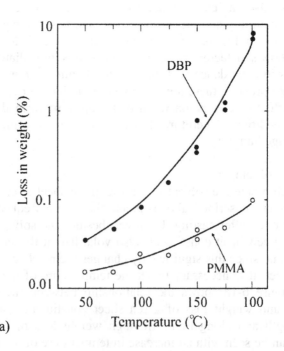

(a)

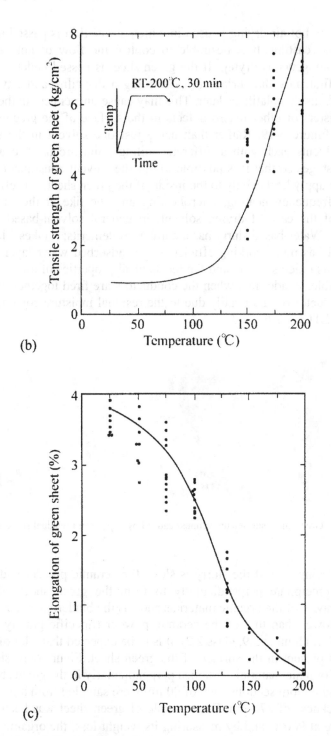

Figure 6-18   The relationship between the heat treatment temperature and weight loss of green sheet constituents, and green sheet tensile strength and elongation.

Typically, hot-air drying is used in which slurry film is passed under the hot air after casting. It is desirable to control the flow of hot air and to achieve homogeneous drying. If the green sheet is passed under the hot air abruptly, first only the surface dries forming a dry film, and unvaporized solvent left inside volatilizes later. This may cause unevenness in the form of cracks, blisters or other macro defects in the surface of the green sheet, as shown in Figure 6-19. Rather than using just one solvent in the slurry, if several solvents each with a different boiling point are used, evaporation occurs in stages so that it is possible to dry the solvent constituents slowly. In order to apply heat evenly to the inside of the green sheet, it is effective to use high frequency heating, microheating and the like at the same time. Because of the ease of drying solvent, in general solvent-based slurry is often used. Water-based slurry has a number of demerits; it takes a long time to dry and has poor working efficiency. An adsorbed water layer tends to remain in its oxide surface, and the mechanical properties of the green sheet are unreliable. In addition, when the conductors are fired together, oxidation of the conductor occurs readily due to the residual moisture content, so that its use for LTCCs is limited.

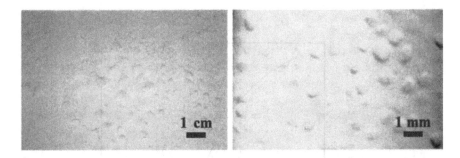

Figure 6-19   Green sheet surface unevenness caused by vapor from residual solvent erupting.

If the drying rate of the slurry is slow, the ceramic powder and organic materials precipitate perpendicularly to form the green sheet. Since the specific gravity of the organic materials such as the binder, plasticizer and so on is far lower than that of the ceramic powder (Specific gravity: organic materials 1.1, alumina 3.9, glass 2.2), it is to be expected that a lot of organic material is present in the surface of the green sheet. Figure 6-20 shows the amount of organic material contained perpendicularly in the green sheet. Top, middle, and bottom sections each of 90 μm were sampled from a green sheet with a thickness of 270 μm. Each sample of green sheet was heated in the atmosphere at 900°C and by measuring its weight loss, the organic material content of each section was calculated. Incidentally, when the raw materials were prepared, 15 wt% of organic material was added. In the results shown in Figure 6-20, the middle part shows an amount of organic material close to that of the original preparation, while the top part is richer in organic material and the bottom part contains less. Figure 6-21 shows cross sections of the microstructure of green sheet close to the top and bottom surfaces. The microstructure in the top and bottom differs significantly, and in the surface part, the presence of a lot of organic material between the ceramic particles can be observed. This difference in perpendicular microstructure becomes more noticeable the thicker the green sheet. It is likely that due to the effect of the difference in drying rate between the top and bottom of the green sheet, differences will occur in the perpendicular density of voids, and as well as the difference in the amount of organic material, this factor is the cause of stress occurring perpendicularly. When there is a big difference in the internal structure of the green sheet between the top and bottom, the sheets may become bowl shaped, and this also has an impact on firing shrinkage behavior.

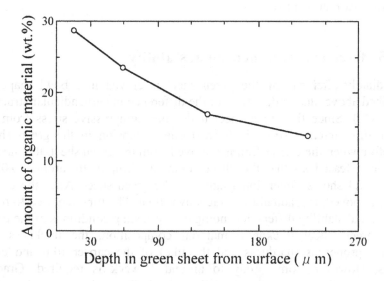

Figure 6-20   The amount of organic material perpendicularly through the green sheet.

When casting the green sheet, there is a plastic carrier film at the bottom so that unlike the free surface at the top, the bottom surface (in contact with the carrier film) has excellent surface smoothness. This has little to do with the perpendicular distribution of the resinous constituents. For this reason, the surface that is in contact with the carrier film is used for printing in the subsequent processes.

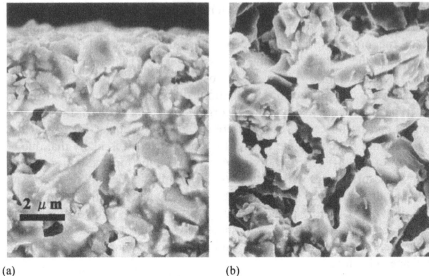

(a)   (b)

Figure 6-21   The microstructure of the top and bottom of a green sheet; (a) the top surface, and (b) near the bottom surface.

6.4. 5   Green sheet dimensional stability

Immediately after casting the green sheet is curved in a bowl shape, as described above, due to the effect of the difference in perpendicular structure (Fig. 6-22). Since the green sheet undergoes compressive stress from the carrier film, internal stress is formed due to tension in the green sheet. Therefore when the carrier film is removed from the green sheet, the internal stress is released so that the sheet contracts abruptly by around 0.05%. Figure 6-23 shows dimensional aging of the green sheet. After the carrier film is removed, it gradually contracts over time. The time required to reach dimensional stability differs depending on the casting conditions of the green sheet, green sheet thickness, material composition, the thickness and physical properties (hardness and the like) of the carrier film and other factors. However from 1 day to around 1 week is required. Gradual relaxation of the stress created by the organic material parts, and gradual dispersal of the residual solvent and moisture content in the sheets are probable causes of the dimensional change.

If the dimensions of the green sheets change after post-processing such as via hole punching, printing and so on, when each green sheet is laminated the positions forming the conductors will not match the sheets above and below causing connection defects between layers. For this reason, after casting the green sheets, it is common to heat them at a low temperature of around 50°C to drive off the solvent and moisture content. In addition, applying minute pressure of around several MPa to the green sheet minimizes the perpendicular distribution of the voids and organic materials which is effective for stabilizing the green sheet. However, since a certain amount of flexibility is necessary for combining each green sheet when laminating them after post-processing, it is desirable to keep the pressure low. After this heating and pressure treatment, it is necessary to let the sheets stand in an environment with a specific temperature and humidity until shrinkage is complete and their dimensions stabilize (aging treatment) (refer to Figure 6-23).

Green sheet

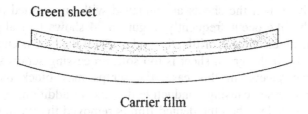

Carrier film

Figure 6-22   Green sheet curved in a bowl shape after casting.

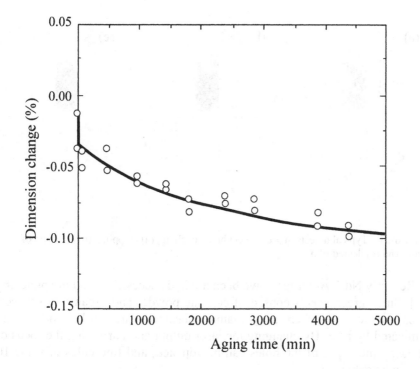

Figure 6-23   Green sheet dimensional aging.

6.5 Via hole punching

In order to form via conductors in the green sheets for connections between layers, it is necessary to form holes with a diameter of around 100 μm in the sheets. For this process the following qualities are required of the green sheets. The dimensions of the sheet must not change during processing, the processing surface must be smooth, processing accuracy must be satisfactory, and process scraps must not stick to the sheets. In general, punching or drilling is used for the process, with NC control. Since there are hard alumina particles mixed into the LTCC green sheets, abrasion of the punch occurs readily. When the sheets are punched with an abraded punch, fine cracks and the like occur frequently. Figure 6-24 shows typical processing defects. If the green sheet is brittle, the underside of the hole is chipped away (a). Conversely, if the green sheet is too soft, processing scraps stick to the bottom of the sheets. In this case, the scraps may block the holes in subsequent processes causing conductive defects. In addition, if the film is pushed into the hole, when the dented film is removed the green sheet itself or conductive powder added in later processes may become embedded causing defects.

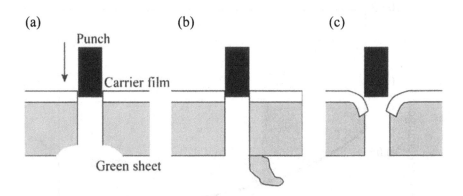

Figure 6-24   Typical defects in green sheet hole punching; (a) chipping, (b) adhesion of scraps, and (c) dented film.

Recently Nd-YAG lasers have been applied successfully to hole punching [25]. Since green sheets consist of ceramic powder and organic substances, holes are formed when the organic substances are broken down and eliminated by heat. By changing the laser output and narrowing the focus of the laser, the depth of the holes can be adjusted, and fine holes of some 10 μm can be achieved.

References

[1] R. B. Runk and M. J. Andrejco, "A Precision Tape Casting Machine for Fabricating Thin Ceramic Tapes", Ceramic Bull., Vol. 54, No.2, (1975) pp. 199-200.

[2] R. E. Mistler, R. B. Runk, and D. J. Shanefield, Ceramic Fabrication Processing Before Firing, Edited by G. Y. Onoda and L. L. Hench, Wiley, New York, (1978), pp. 411-48.

[3] K. Saito, "Use of Organic Materials for Ceramic-Modeling Process-Binder, Deffloculant, Plasticizer, Lubricant, Solvent, Protective Colloid-," J. of the Adhesion Society of Jpn, Vol. 17, No. 3, (1981), pp. 104-113.

[3] E. P. Hyatt, "Continuous Tape Casting for Small Volumes," Ceramic Bull., Vol. 68, No. 4, (1989), pp. 869-70.

[4] Y. Fujii Slot-Die Method, Subject and Solution of Manufacturing Process of Ceramics for Microwave Electronic Component, Technical Information Institute, (2002), pp. 73-84.

[5] E. B. Gutoff and E. D. Cohen, Coating and Drying Defects, John Wiley & Sons Inc. (1995).

[6] T. Yamano, Research & Development of Ceramic Devices & Material for Electronics, CMC, Tokyo, (2000), pp. 55-62.

[7] Y. T. Chou, Y. T. Ko and M. F. Yan, "Fluid Flow Model for Ceramic Tape Casting," J. Am. Ceram. Soc. Vol. 70, No. 10, (1987), pp. C280-82.

[8] Pitchumani and V. M. Karbhari, "Generalized Fluid Flow Model for Ceramic Tape Coating," J. Am. Ceram. Soc., Vol. 78, No. 9, (1995) pp. 2497.

[9] K. Otsuka, Y. Osawa and K. Yamada, "Studies on Conditions in Casting Alumina Ceramics by the Doctor-Blade Method and their Effects on the properties of Green Sheets (Part 1)" J. Ceram. Soc. Jpn., Vol. 94, No. 3, (1986) , pp. 351-359.

[10] K. Otsuka, W. Kitamura, Y. Osawa and M. Sekibata, "Studies on Conditions in Casting Alumina Ceramics by the Doctor-Blade Method and their Effects on the properties of Green Sheets (Part 2)" J. Ceram. Soc. Jpn., Vol. 94, No. 11, (1986), pp. 1136-1141.

[11] K. Hirao, "Microstructural Control of Silicon Nitrides for High Thermal Conductivity," Bull. Ceram. Soc. Jpn., Vol. 33, No.4, (1998), pp. 276-280.

[12] S. Otomo, M. Kato and K. Korekawa, "Firing Shrinkage of Alumina Green Sheet," J. Ceram. Soc. Jpn., Vol. 94, No. 2, (1986), pp. 261-266.

[13] Japan Industrial Standard, Surface Roughness, JIS B0601

[14] H. Sekiguchi, H. Takeyama, R. Murata, and H. Matsuzaki, Transactions of the Japan Society of Mechanical Engineers, Vol. 43, No. 1, (1977), pp. 374.

[15] Japan Industrial Standard, Testing method for tensile properties of plastics, JIS K7113

[16] K. Kubo, Y. Nakagawa, E. Mito and S. Hayakawa, Powders (Theory and Application), Maruzen, (1962), pp. 161-162.

[17] K. Otsuka, and T. Usami, "Considerations regarding interaction between the burn out of polyvinyl butyral binder and sintering of alumina ceramics in a reducing atmosphere (Part 1)," J. Ceram. Soc. Jpn., Vol. 86, No. 6, (1981), pp. 309-318.

[18] F. Carli and A. Motta, "Particle Size and Area Distributions of Pharmaceutical Powders by Microcomputerized Mercury Porosimetry," J. Pharmaceutical Science, Vol. 73, No. 2, (1984), pp. 191-202.

[19] J. V. Brakel, S. Modry, and M. Svata, "Mercury Porosimetry: State of the Art, " Powder Technology, Vol. 29, (1981), pp. 1-12.

[20] R. A. Gardner and R. W. Nufer, "Properties of Multilayer Ceramic Green Sheets," Solid State Technology, May (1974), pp. 38-43.

[21] J. S. Reed, T. Carbone, C. Scott, S. Lukasiewicz, Processing of Crystalline Ceramics, Plenum, New York, (1978), pp. 171.

[22] M. Sasaki, "Binders for Tape Casting Method," Bull. Ceram. Soc. Jpn., Vol. 32, No. 10, (1997), pp. 812-818.

[23] R. E. Becker and W. R. Cannon, "Source of Water and Its Effect on Tape Casting Barium Titanate," J. Am. Ceram. Soc. Vol. 73, No.5, (1990), pp. 1312-17.

[24] N. Watanabe and N. Kamehara, Bull. Ceram. Soc. Jpn., Vol. 32, No. 9, (1997), pp. 762-765.

[25] M. Fujimoto, S. Sekiguchi, H. Takahashi, M. Nakazawa, and N. Narita, "New Laser Processing for Ceramic Multilayer Components and Modules," J. Ceram. Soc. of Jpn, Vol. 108, No. 9, (2000), pp. 807-812.

Chapter 7
Printing and laminating

In the previous processes, the green sheets were formed from slurry using a doctor blade, and via holes were formed in the green sheet. This chapter covers the processes before firing; via filling, printing, and laminating. Via filling is the process in which the via holes formed in the green sheet for conduction between layers are filled with conductive powder, paste or the like, and in the subsequent printing process, a wiring conductor pattern is formed in alignment with these vias with the screen printing technique. Additionally, the multiple layers of green sheet on which vias and wiring are formed are aligned. These green sheets then undergo thermocompression bonding to combine them into one unit to form a green sheet laminated body. When producing LTCC circuit boards, discrete wiring is formed in the z axis and x and y axes in the former two processes, while in the lamination process, the pattern for the three dimensional circuit wiring network is formed. Since the quality of the conductor formed in these processes has a significant impact on the characteristics of the sintered conductor obtained after the subsequent firing process, the process technology must be controlled carefully.

The sections in this chapter for via filling and printing are technologies typically used for common LTCC products. On the other hand, the section on lamination describes the process technologies for forming multilayered structures of ten layers or more. However, this is a very useful technology for achieving a high quality and reliable multilayered structure with wide manufacturing margins, even when making laminated circuit boards with few layers.

7.1 Printing

In general, the screen printing technique known as gap printing is used for printing the wiring pattern on the green sheet. Gap printing is a method in which a gap is set between the mask and the green sheet and when the squeegee passes over the mask, the conductive paste is pushed through the openings in the mask onto the green sheet. With this method, stress is applied to the conductive paste by the squeegee, the tip of which undergoes elastic deformation. The squeegee pushes the paste through the openings of the mask so that the paste is applied to the green sheet. At the same time, the screen is released from contact with the paste. For this reason, achieving a regular amount and shape of deformation in the tip of the squeegee is a key point for maintaining print quality [1].

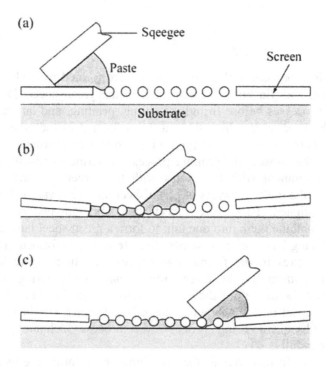

Figure 7-1   Screen printing process.

Printing the conductor with excellent print quality as shown in Figure 7-2, without any bleeding or blurring, is important. Good print quality means in the right position (printing precision), in the right amount (transfer ratio), in the right shape (transfer shape), and with stable printing (repeatability). In order to achieve this, the following 4 items must be optimized; 1. screen specifications, 2. printing process conditions, 3. paste characteristics, and 4. green sheet characteristics. The details of each item are shown below.

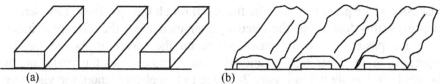

Figure 7-2   Typical defects in printed patterns; (a) no bleeding, (b) bleeding.

7.1.1   Screen printing screen specifications

The screens used for thick film printing of conductive paste for LTCCs are a mesh structure woven from fine stainless steel wire, and the parts other than the openings are covered in emulsion. It is necessary to optimize the diameter of the wire for the mesh, the number of strands in the mesh, and the thickness of the emulsion in accordance with the purpose of printing. For example, when making a thick film, a screen using thick wire mesh is used, with thick emulsion. When thick wire is used, each opening hole is large, although inevitably the aperture ratio is smaller. Since a difference in level is formed where the wires cross, unevenness in the surface of the film deposited occurs readily. In addition, when printing a fine pattern with narrow line widths, it is effective to use a screen made of wire with a small diameter for the mesh so that the pattern has a large aperture ratio, with thin emulsion.

Furthermore, since the tension of the screen becomes slack with repeated use, it is necessary to check it periodically. Also, the screen must be washed adequately in order to form patterns with good repeatability. In particular, the triangular openings at the boundary of the aperture pattern and the edge of the parts coated in emulsion are particularly prone to blocking with paste residue.

7.1.2   Printing process conditions

In the printing process it is desirable for the pattern formed on the screen to be transferred directly onto the green sheet in the medium of the paste. The printing conditions have the greatest impact on this print transfer quality, although the appropriate printing conditions vary depending on the screen and paste used so that they must be optimized on each occasion. The next section covers the various process parameters.

(1) Squeegee speed
Squeegee speed is decided by the printing time and viscosity of the paste when printing. In general, reducing squeegee speed increases the printing

time and improves printability. However, when squeegee speed is fast, the viscosity of the paste falls, and its fluidity through the openings in the screen improves (this depends on the viscosity characteristics – the thixotropy index – of the paste). When the paste placed on the screen is moved by the squeegee to the openings, rotational force is applied to the paste and it begins to roll as shown in Figure 7-3. Due to this phenomenon, the viscosity of the paste itself falls so that it can be pushed through the openings more easily, improving the transfer quality of the paste. In other words, print quality can be improved by causing the paste to roll thereby reducing its viscosity, then moving the squeegee at a comparatively low speed [3].

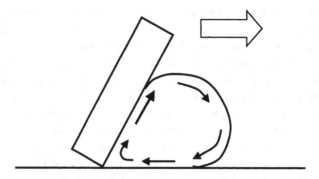

Figure 7-3   The paste rolling phenomenon.

(2) Squeegee pressure

In the process in which the paste is transferred to the substrate (green sheet), the squeegee presses down the screen, and it is important for the screen to come in contact with the green sheet positively. It is also important for the screen to separate from the green sheet positively in the locations where the squeegee passes.

The most appropriate pressure depends on the amount of deformation of the screen itself and the gap between the screen and substrate (green sheet). The amount of deformation varies according to the diameter of the wire used for the screen and its tension, and the printing gap must be changed depending on the screen used. The amount of gap should be determined taking into account the force of the screen returning from the flexed position when the downward pressure is released, and the snap off properties of the screen. Then once the appropriate squeegee pressure is set so that when the screen is pushed down it comes in contact with the green sheet, it is easy to optimize the squeegee pressure conditions

In general, high squeegee pressure causes bleeding readily, while conversely, low pressure may result in blurring. It is also effective to use a soft squeegee material in order to increase the margins of the appropriate printing conditions.

(3) Squeegee angle
If good printing cannot be achieved by adjusting the squeegee speed and squeegee pressure described above, it is also possible to ameliorate defects such as bleeding and blurring by adjusting the squeegee angle. Adjusting the angle is especially effective when the paste does not roll readily.

7.1.3 Paste characteristics

The term 'paste' indicates an intermediate property between a solid and a liquid, being both elastic and viscous (showing viscoelasticity). Rheological handling is necessary for paste [4].

In a system where paste fills a narrow gap between two parallel plates, when only one plate is moved in parallel at shear rate D, the shearing stress τ required to move the plate is expressed with the following equation. Poise or pascal seconds from the SI system of units are used as the unit for viscosity η, and the following relationship obtains [5].

$\eta = \tau/D$ (1 P = 100 cP, 1 Pa $\cdot$ s = 10 P, 1 mPa $\cdot$ s = 1 cP)

Figure 7-4 shows the viscous behavior of a representative fluid. Among fluids there are those that have yield values. This kind of fluid does not begin to flow even when force is applied, until the yield value τ_0 is exceeded. Conductive paste exhibits fluid behavior close to that of pseudoplastic fluids. The viscosity of this kind of fluid shows a tendency to fall as the shear rate increases. As shown in Figure 7-5, when the relationship between viscosity and shear rate are plotted double logarithmically, the results is a more or less straight line. In the figure, when the shear rate is increased from D_1 to D_2, viscosity is reduced from $\eta 1$ to $\eta 2$, and the rate of change of the viscosity with regard to the shear rate is expressed as

$$\frac{\log(\eta_1/\eta_2)}{\log(D_2/D_1)}$$

This value is defined as the TI (thixotropy index) which is used as an index for showing the degree of shear rate dependence of conductive paste viscosity.

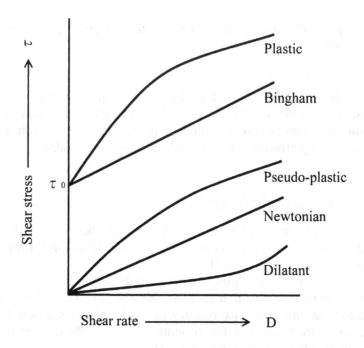

Figure 7-4   The viscous behavior of various kinds of fluid.

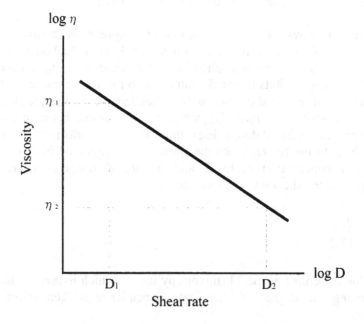

Figure 7-5   The relationship between shear rate and viscosity (TI - thixotropy index).

If it is stored for a long period of time, conductive paste hardens, but when shearing stress is applied again and it is stirred, its fluidity increases. During processing too, the phenomenon occurs in which the paste softens when it undergoes strong shearing stress on the screen, but when left standing its viscosity returns. This phenomenon is called thixotropy. When the paste undergoes strong shearing stress, the vehicle layer of the particle surface is broken down, and the structure forming capability due to the cohesion between particles falls so that the paste softens. When left to stand, the vehicle layer recovers and viscosity returns. This breakdown of the structure by stirring and recovery of the structure when resting is a function of time, and recovery time differs depending on the type of paste. The recovery rate for viscosity is generally obtained as shown in Figure 7-6. First measure the viscosity until the rotational frequency of the viscosimeter changes from 3 rpm to 30 rpm. Taking 10 rpm as standard viscosity (η_a), after measuring viscosity at 30 rpm, reduce the rotational frequency to 10 rpm and after 1 minute read off the viscosity (η_b) and find the recovery rate for viscosity with the following equation. A lower value for R indicates faster recovery.

$$R = \frac{\eta_a - \eta_b}{\eta_a} \times 100$$

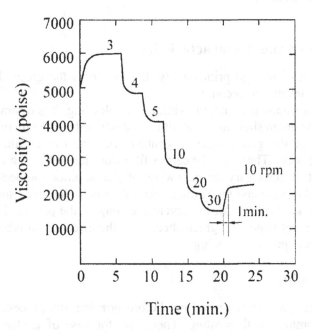

Figure 7-6 The method for finding the recovery rate for viscosity.

The two indices above (TI and R value) are useful for quality control of the paste. In order to form fine wiring patterns that are not sloppy and have good thickness (refer to Figure 7-7), it is desirable for the fluidity of the paste to increase when printing with a high shear rate, and for the paste to have high viscosity after printing (with a low shear rate from the weight of the paste itself), in order to maintain its shape without running. To vary these numeric values, it is necessary to change the composition of the inorganic material ingredients in the paste, the shape of the powder particles used, the composition of the organic material ingredients and so on.

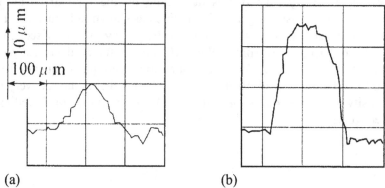

(a)   (b)

Figure 7-7   Cross section of printed conductor pattern (a) with significant running, (b) without significant running.

7.1.4 Green sheet characteristics

In order to achieve high print quality, the substrate, the green sheet itself, must also be taken into account.

To form a fine wiring pattern without any bleeding, it is desirable for the surface of the green sheet to be smooth. As described in the previous section, the surface of the green sheet in contact with the carrier film has little surface roughness. Therefore the carrier film side should be used for printing. However, it is necessary to be aware of the solution properties of the solvents in the paste and the resinous constituents in the green sheet, and to optimize the kind of solvent to prevent bleeding of the pattern. In addition, the voids present inside the green sheet have the effect of absorbing solvent which works to prevent bleeding.

7.2 Via filling

Although this process is carried out before printing, the process has many points in common with printing. Therefore, for ease of explanation, it is presented after the section on printing. Via filling is the process in which the via holes formed in the green sheet (using a punch, drill, laser [6, 7] and so on) are filled with the conductor. In general, a squeegee is used to fill the holes with the conductor. It is desirable to fill the via holes in the green sheet uniformly with the conductor. Filling should not be as dense as possible. Rather, the optimal filling ratio for the conductor depends on the packing density of the powder in the surrounding ceramic green sheet. If conductor is packed in more densely than the packing density of the powder in the green sheet, the volume of the conductor becomes larger than the volume inside the hole after the subsequent firing process. Since this causes compressive stress on the conductor and tensile stress on the ceramic, distortion of the via hole interface occurs, resulting in reduced reliability of the via. Conversely, an insufficient filling ratio for the via conductor causes open circuits in the via conductors.

Although the conductor is formed using a squeegee similarly to printing, the object of the process is completely different. While printing applies the conductor onto a flat surface, in the via filling process, conductor is packed into cylindrical via holes that have a high aspect ratio, so that the fluidity of the filling is important. Paste or powder is generally used as filling.

The problem with paste filling is that the solvent in the paste dissolves the binder in the green sheet so that the via hole spreads and bleeding can be seen. In addition, if the via hole can actually be filled completely with paste, since the amount of solid ingredients in the paste is around 70% (depending on the paste), the actual filling ratio of the conductor is limited. However, as paste has good fluidity, it offers the benefits of being easy to pack in while it has good workability.

When powder is used as filling, it is difficult to fill the via holes since the fluidity of copper powder itself is poor. For this reason, cavities like those shown in Figure 7-8 (a) occur readily inside the via. The occurrence of this kind of cavity can be reduced by vibrating the green sheet. In addition, since there is no lytic reaction and so on between the green sheet and the filling, adherence to the green sheet is poor and the powder can easily fall out of the hole. In order to prevent the powder falling out, the method shown in Figure 7-8 (b) can be used in which a cap is first formed with paste at the bottom of the via hole before it is filled with powder. It is also possible to achieve a higher filling ratio than with the paste method by applying pressure after filling with powder.

It is necessary to select a powder that has a low angle of repose and excellent fluidity. Additionally, in order to increase the filling ratio, it is effective to use a powder with a mixture of 2 or more powders with varying particle sizes.

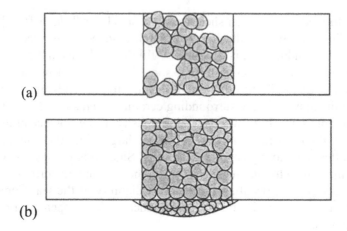

Figure 7-8   Schema of via filling; (a) cavities in conductive powder in the via hole, and (b) process in which a cap is first formed with paste at the bottom of the via hole before it is filled with powder.

7.3 Laminating

The purpose of the lamination process is to make a single substrate by aligning several layers of green sheet on which vias and wiring have been formed. Pressure and heat is then applied and the several individual green sheets are combined into one unit. Alignment is a process required to ensure that the circuit contacts in the three dimensional network connect properly. The purpose of the stacking process is to achieve a single macro focus of contraction when fired. This is important for increasing the mechanical reliability of the fired substrate. Each green sheet must bond sufficiently with the green sheets above and below it.

7.3.1 Laminating process technologies

(1) Alignment
Each green sheet is aligned using an aligner based on the principle shown in Figure 7-9. As shown in the Figure, the equipment consists of 4 CCD cameras, an xy-θ stage, and a fixed stage.

The following procedure is used for alignment. (a) The green sheet is placed on the xy-θ stage located between the CCD cameras and the fixed stage. As the markers formed at the 4 corners of the green sheet do not match up with the center of the cameras, the image shown in the figure appears in the camera. (b) In order to align the green sheet markers with the central axes of the cameras, the position of the cameras is shifted. Then the camera position is fixed as the standard optical axis for the alignment operation. (c) The green sheet placed on the xy-θ stage is moved in parallel to the optical axis and is placed on the fixed stage. (d) The second green sheet is placed on the xy-θ stage and by moving or rotating the stage, the center of each camera is matched up with the green sheet markers. If the green sheet markers and the center of the cameras do not match up at this step, it is generally the result of the green sheet stretching during handling, melting reactions between the solvent in the paste and the resin in the green sheet during printing, or slight shrinkage of the green sheet. Shrinkage of the green sheet after printing in particular varies in significance depending on the pattern density and pattern thickness. The dimensions of each green sheet must be checked, and green sheets with a significant amount of shrinkage should be removed from the process as defective articles. In addition, as described in the section on printing, the ceramic particles within the green sheet may be oriented in the direction of casting so that after firing, the amount of shrinkage varies in the x-y axis. In order to reduce the difference in amount of firing shrinkage in the x-y axis over the whole substrate, it is common to stack each layer of green sheet at 90° to the direction of casting (refer to Figure 7-10 (a)). (e) The green sheet aligned on the xy-θ stage is next moved to the fixed stage. Then by repeating the alignment in (d) and (e), the layers are built up. (f) Finally, the stack of aligned and layered green sheets is taken out and placed in a mold.

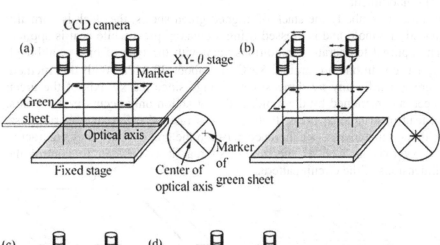

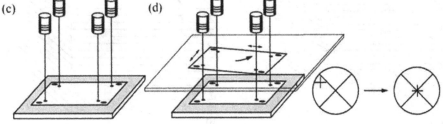

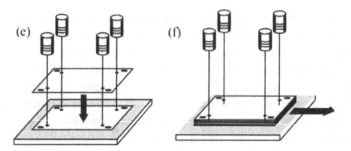

Figure 7-9   The general principle of a green sheet aligner.

If a mark is made on the edge of each green sheet and this edge is cut after laminating, the deviation in the marks can be observed and the alignment accuracy in the z axis can be checked before firing (refer to Figure 7-10 (b)). In addition, since there is irregularity in the perpendicular structure of the green sheet, the bonding strength between layers during stacking and the direction of shrinkage when firing may differ depending on the surface. When placing the green sheets, it is important to arrange the surfaces of the sheets in order (for example, always placing the carrier surface face up).

(2) Embodiment
In order to embody the stack of aligned green sheets, the stack is normally placed in a mold and is pressed using a uniaxial press while heat is applied. The optimal temperature conditions vary with the type of binder, although typical conditions are around 80°C and about 30 MPa. With this method, there is hardly any xy dimensional change since the xy axis of the green sheet is constrained by the mold, and contraction only occurs in the z axis. Instead of a uniaxial press, there is also a method using an isobaric press. Although this increases adherence between the green sheets, they contract in the xy axis as well as the z axis so that it is difficult to control the dimensions of the circuit pattern.

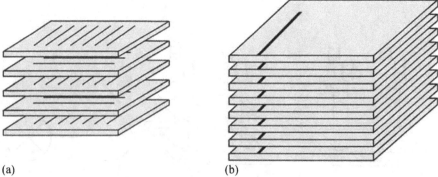

(a)   (b)
Figure 7-10   Methods of placing the green sheets; (a) placed rotated at 90° and, (b) z axis alignment check.

Besides this one-time lamination method, there is a build-up method of lamination.

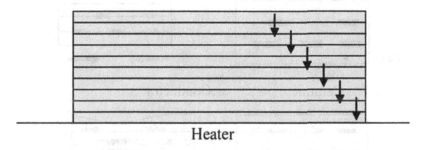

Figure 7-11   Laminating with the build-up method.

In this method, after alignment the green sheets are laminated layer by layer on a hot platen and pressure is applied. This is efficient since alignment and lamination are performed at the same time. However, as shown in Figure 7-11, the lowest green sheet is constantly heated, and it is pressed cumulatively many times. Therefore the lowest green sheet looses its flexibility so that the structure of the laminated body in the z axis becomes uneven making the method unsuitable for substrates with many layers.

Whichever method is used, after lamination, the more layers there are, the greater the difference in thickness becomes between parts containing conductor and parts with ceramic only. Since the thickness of the conductive paste is around 10 μm, when around 10 layers are laminated, a difference equivalent to the thickness of 1 layer of green sheet arises. In extreme cases, this phenomenon causes the delamination explained below. Furthermore, before firing the stacked laminated body, cutting the edges slightly is effective in stabilizing firing shrinkage. During firing, if the edges catch on the setter, friction with the setter occurs so that isotropic contraction cannot be obtained.

7.3.2   Faults arising in the laminating process

The most significant fault that occurs in the laminating process is delamination. If there are defects such as poor bonding between layers in the laminated body, these defect locations can become the starting point for delamination between layers. This is due to the fact that, when there is poor bonding between layers in the laminated body, each individual green sheet becomes separate and contracts, and there is no central point where contraction is focused (Figure 7-12). Delamination is not only caused by the laminating process; it also occurs in the firing process (for example, due to firing mismatch and the like) although this is covered in the chapter on firing.

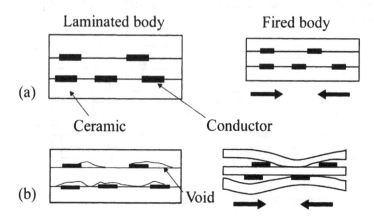

Figure 7-12   Conceptual diagram of delamination; (a) with sufficient embodiment of the laminated body, and (b) with insufficient embodiment of the laminated body where each individual green sheet becomes separate and contracts.

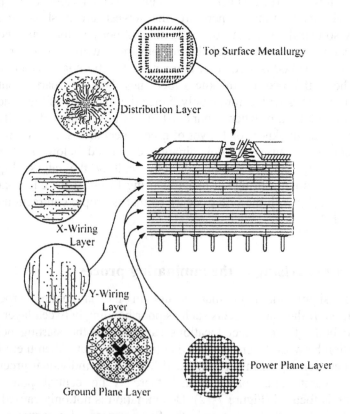

Figure 7-13   The conductor pattern in a multilayer ceramic circuit board for use in a mainframe computer (Ref. [8]).

This kind of delamination is a phenomenon that occurs particularly with multilayer substrates of 10 layers or more in which the effect of the conductor becomes more significant. Figure 7-13 shows the conductor pattern in a multilayer ceramic circuit board for use in a main-frame computer [8]. As the figure makes clear, compared with the wiring conductor pattern, the power supply and ground patterns occupy a larger area so that when the sheets are laminated, the bonding area between green sheets is less (there is more conductor than green sheet). As explained earlier, in the parts where conductor is formed and where it is not, the total thickness of the z axis of the laminated body differs.

Adherence between layers of green sheet in the laminated body is achieved with the following two methods.

- By the bonding of resinous constituents melted by the heat
- By the mechanical bonding of unevenness in the joint surface (interlock bonding)

Since the resinous constituents at the interface between green sheets are the same, they have good bond strength. In order to achieve high strength at the interface between conductor and green sheet, it is essential to consider the compatibility of the resinous constituents used in the conductor and those in the green sheet. In addition, to increase the bond strength of interlock bonding, the green sheet itself must be soft and pliable. For this reason, it is best if there is a certain amount of voids in the green sheet.

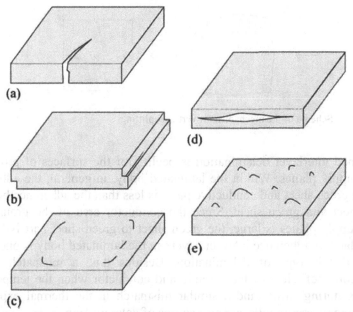

Figure 7-14   Typical forms of delamination; (a) vertical splitting, (b) stepped interlayer delamination, (c) circular delamination, (d) internal interlayer delamination, and (e) surface blistering.

Figure 7-14 shows typical delamination defects seen after firing. The details of each type of delamination and its causes are explained below.

Vertical splitting is delamination in which a crack forms from the middle of the edge of the substrate towards its center, bisecting the substrate. Adhesion between layers is strong, and no delamination is observed between layers. During lamination, deforming forces are concentrated in the center of each edge of the green sheet. However, as the green sheets are enclosed in the mold, no distortion in the x and y axes is possible. As shown in Figure 7-15, it is likely that since the density of the green sheets is higher at the center of the edges of the green sheet, the cracks form during firing in the center of the edges. This kind of delamination is also caused by uneven pressure on certain areas during lamination due to poor parallelism in the press or mold.

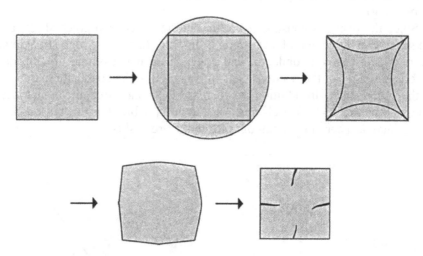

Figure 7-15   Schema of the mechanism of vertical splitting.

Stepped interlayer delamination is peeling of the surfaces of ground or power supply planes. Within the laminated body, in general, the adherence between green sheet and conductive paste is less than the adherence between green sheet and green sheet. Since the conductor area of the ground and power supply planes is large, the green sheet to green sheet part is limited, and the lack of adherence between layers in the laminated body is one of the causes of this type of delamination. Besides this, a mismatch in the contraction coefficient of the ceramic and conductor when the temperature increases during firing, and a similar mismatch in the thermal expansion when the temperature falls, are also causes of delamination.

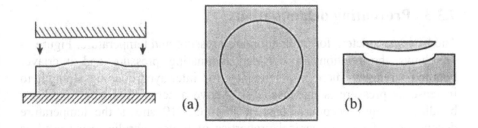

Figure 7-16 Schema of the mechanism of circular delamination; (a) view from above, and (b) cross section.

Circular delamination is where a circular portion only peels off from the fired body as shown in Figure 7-16, and in extreme cases, a plate shaped piece delaminates. This probably occurs due to the fact that normally, since the conductor pattern is concentrated in the center of the green sheet so that the total thickness of the middle part only is greater and is raised higher, uneven pressure is applied during lamination.

Internal interlayer delamination is where delamination occurs at the interface between the internal conductor and the ceramic, although the cracks do not reach to the outside of the substrate. The cause of the problem is an area of poor adherence between the green sheet and conductive paste in the laminated body. When there are many layers, or when the conductor on each layer is thick, there is a great difference in the thickness of the parts including the conductor and the parts with ceramic only, so that the laminated body is like a sandwich with a lot of filling (refer to Figure 7-17). Since this delamination occurs in order to release the stress within the laminated body, it takes a similar form to stepped interlayer delamination and circular delamination.

With surface blistering, there is balloon shaped expansion close to the surface of the substrate after firing. This is formed when undissolved matter in the binder and gas dissolved in the raw glass material is released at high temperature during firing. It is important to allow gas to be released from the laminated body sufficiently before sintering of the surface parts of the substrate takes place (while pores from which the internal gasses can be released are present).

Figure 7-17 Schema of the mechanism of internal interlayer delamination.

7.3.3 Preventing delamination

The basic parameters for lamination are pressure and temperature. Figure 7-18 shows the relationship between laminating pressure and interlayer bonding strength. There is a tendency for interlayer bonding strength to increase as pressure is increased [9]. Above a certain pressure, interlayer bonding strength becomes constant. Figure 7-19 shows the temperature dependency of the mechanical properties of typical acrylic resins used as binder and so on. Tensile strength tends to fall when temperature rises. In addition, the maximum values for elongation can be seen in 2 temperature ranges [10]. Figure 7-20 shows an investigation of the laminated density when the laminating temperature and pressure are changed. Although the maximum points do not match exactly, a tendency can be seen for laminating density to be highest in the 2 temperature ranges around 50°C and 175°C in a high pressure range. In other words, it is likely that the elongation that results from fluidity of the binder resin at high temperature has a significant impact on the characteristics of the laminated body. Figure 7-21 shows the microstructure between layers and within layers of the laminated body taken with a SEM. Between layers, the binder resin can be seen to be concentrated. It is likely that the resin is responsible for bonding between layers, so that it is necessary to take sufficient account of the fluidity of the resin and optimize the laminating temperature and pressure.

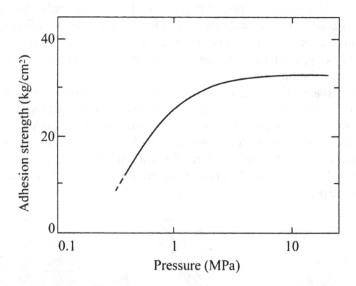

Figure 7-18   The relationship between laminating density and the interlayer bonding strength of the laminated body.

Figure 7-22 shows a cross section of a laminated body after tensile stress is applied to the surface and vertical axis of the laminated body. A crack runs through the conductor, and delamination can be seen. Between layers, the parts with conductor can be considered to be the weakest parts.

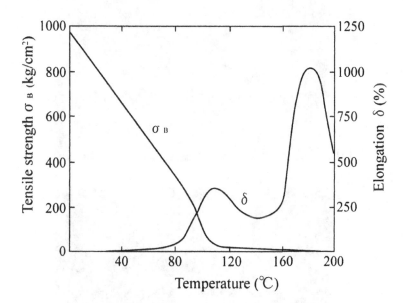

Figure 7-19   The temperature dependency of the mechanical properties (tensile strength and elongation) of acrylic resin.

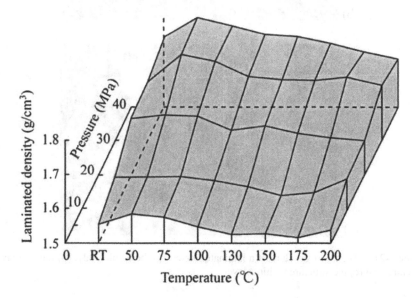

Figure 7-20   The relationship between laminating conditions (temperature and pressure) and interlayer bonding strength (using acrylic resin as the binder).

Figure 7-21   The microstructure of a laminated body; (a) the whole body, (b) the interlayer structure, and (c) the structure within a layer.

In order to prevent the delamination explained above, it is effective to optimize the laminating conditions (temperature and pressure), improve the adhesive properties of the resinous constituent of the green sheets and conductive paste, and to check the parallelism of the press and so on. In addition, forming dummy conductors in the edge parts where conductors are not normally formed to adjust their thickness is an effective method of preventing delamination.

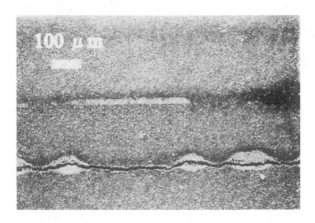

Figure 7-22 The cross sectional structure of a laminated body when tensile stress is applied.

References

[1] J. Savage, Handbook of Thick Film Technology," P. J. Holmes and R. G. Loasby eds., Electrochemical Publications Ltd., Scotlamd (1976).

[2] J. Bradigan, "Recommendations for Screen Printing Fine Line Images in Circuitry," Insulation/Circuits, Jan. (1980), pp. 33-38.

[3] Fine Screen Printing Technology, Technical Information Institute (2001).

[4] E. Hirai, Rheology for Chemical Engineers, Science Technology, Tokyo, (1978)

[5] J.R. van Wazer, Viscosity and Flow Measurement, Interscience Publishers, John Wiley & Sons (1966).

[6] M. Fujimoto, N. Narita, H. Takahashi, M. Nakazawa, Y. Kamiyama, and S. Sekiguchi, "Miniaturization of Chip Inductors using Multilayer Technology and Its Application as Chip Components for High Frequency Power Modules," Industrial Ceramics, 12, (2001), pp. 26-28.

[7] W. W. Koste, "Electron Beam Processing of Interconnection Structures in Multilayer Ceramic Modules," Metall. Trans., Vol. 2, No. 3, (1971), pp. 729.

[8] C. W. Ho, D. A. Chance, C. H. Bajorek, and R. E. Acosta, "The Thin-Film Module for High Performance Semiconductor Package", IBM. J. Res. Develop., Vol. 26, No. 3, May (1982), pp. 286-296.

[9] R. A. Gardner and R. W. Nufer, "Properties of Multilayer Ceramic Green Sheets," Solid State Technology, May, (1974), pp. 38-43.

[10] Y. Oyanagi, Introduction to Polymer Processing Rheology, Agune, Tokyo (1996).

Chapter 8

Cofiring

In the previous chapter, each green sheet on which conductor wiring and vias are formed was first aligned, then stacked and embodied to form the circuit board before firing. This chapter explains the process in which the laminated body is heated at high temperature so that the conductor and the ceramic in the laminated body are fired at the same time. When the firing process is complete, the outside edges of the substrate are cut. In some cases the surface of the substrate is polished and a thin film multilayer structure is formed on the fired substrate by forming thin film conductors with the photolithographic process, and polyimide layers using spin-coating. This thin film process uses technologies in common with the Multi-Chip-Module (MCM-D) manufacturing process and it is not peculiar to LTCCs. Therefore in this book, the cofiring process is treated as the final LTCC process.

The sintering behavior of the ceramic in the LTCC is covered in Chapter 2. In addition, the conductive material and conductive paste are described in part in Chapter 3. This chapter focuses on the situation when copper is used as the conductive material, and considers the sinterability of the copper, and co-sintering properties of the copper and ceramic.

In the LTCC cofiring process, normally the organic constituents such as binder and so on are driven off and eliminated in the dewaxing process before each material is sintered and densified. The difficulty of firing a ceramic substrate using copper as the wiring lies in firing the dissimilar materials – copper and ceramic – at the same time. The following 3 items are the most important technical points.

(1) Controlling the firing shrinkage and its variation in the substrate as a whole
(2) Controlling the firing shrinkage behavior of both materials to keep them free of micro and macro defects
(3) Achieving both antioxidation of the conductor metal and elimination of the binder during the firing process

For example, if techniques (1) and (2) cannot be controlled, the substrates bend or develop waves as shown in Figure 8-1. Furthermore, it is necessary to conduct technology development in order to achieve carbon residues in the substrate of less than 100 ppm after firing [1]. Figure 8-2 shows the relationship between the amount of carbon residue in the ceramic substrate and the withstand voltage of the ceramic. Since carbon is conductive, any carbon remaining in the ceramic drastically reduces its insulating properties. Of these points, item (3) can be considered to be an issue unique to using copper as a conductor. The details of each of the 3 points are covered below.

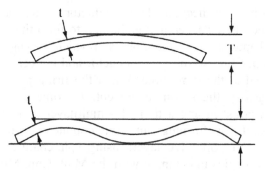

Figure 8-1   Conceptual diagram of bends and waves in the substrate (bend and waves = T - t).

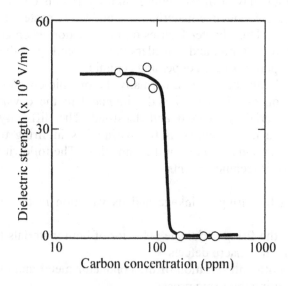

Figure 8-2   The relationship between the amount of carbon residue in the ceramic substrate and the withstand voltage.

8.1 Sintering the copper

According to the thermodynamic equilibrium diagram in Chapter 3, Figure 3-6, silver oxide and palladium oxide deoxidize to a metallic state at close to 200°C and 850°C respectively, even in a 100% oxygen atmosphere. For this reason, it is easy to sinter silver and palladium at high temperature without them oxidizing. On the other hand, copper oxides are stable up to high temperatures at air atmosphere (oxygen partial pressure: 0.2 atm) so that copper can be expected to oxidize easily. Therefore with copper, operations such as deoxidizing the firing environment and so on are necessary, and controlling the sintering of copper is more difficult than that of silver and palladium.

Figure 8-3 shows the microstructure of copper paste when fired at various temperatures. Sintering is the chemical bonding formation of particles at high temperature causing atomic diffusion between particles. Although the particles in the compact are in point contact before sintering, when heated the surface of the particles coalesces and the particles become more closely integrated (the surface energy of the particles tends to be small). In the figure, growth of the particles can be observed as the firing temperature increases. At a firing temperature of 700°C the particle size of raw powder is around 3 μm although it grows to around 10 μm at 900°C. Furthermore, with firing at 900°C or more, it can be observed that closed pores are formed. There is a tendency for the diameter and amount of these pores to increase as the temperature rises. There are three possible causes for the formation of these pores. One is the copper vapor formed with the increase in the vapor pressure of the copper. The second is the oxygen desorbed from the layer of oxides formed at the particle surface due to deoxidation. The third is possibly due to gas from the decomposition at high temperature of organic resin contaminants in the paste that were not decomposed in the low temperature range.

Figure 8-4 shows the firing shrinkage behavior of copper paste. As shown in the figure, expansion is frequently observed at high temperatures of 600°C or more. This is likely to be due to the occurrence of the internal pores noted above, and volume expansion due to oxidation of the copper. Although the firing of the copper in LTCCs often takes place in a nitrogen atmosphere, it is important to ensure control of the oxygen concentration since oxygen adulterant in the atmosphere promotes oxidation.

8.2 Controlling firing shrinkage

In order to control the firing shrinkage rate of LTCCs which use copper conductor, it is necessary to consider the firing shrinkage rate of the ceramic as a simple substance, as well as the firing shrinkage rate of the copper wiring.

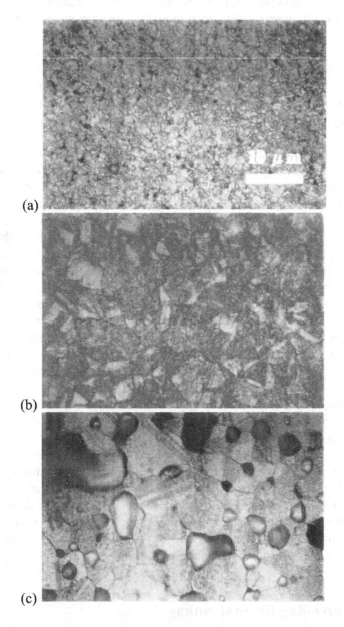

(a)

(b)

(c)

Figure 8-3   The microstructure of copper paste when fired at various temperatures: Firing temperature (a) 700°C, (b) 900°C, and (c) 1,050°C.

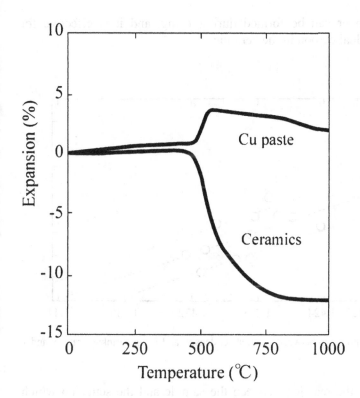

Figure 8-4   The firing shrinkage behavior of copper paste.

(1) Ceramic

The firing shrinkage rate of the ceramic itself is significantly affected by the characterization of the glass base powder and the characteristics of the green sheet. By sorting the powder and using powder with a uniform particle size, it is possible to reduce variation in firing shrinkage rates. There is a strong correlation between green sheet density and its firing shrinkage rate, shown in Figure 8-5. In this figure, in order to keep the firing shrinkage rate between 16.4% and 16.5%, it is necessary to control green sheet density between 1.421 and 1.424 g/cm^3, so that the margin for green sheet density is extremely small.

As a means of stabilizing contraction coefficient, the use of two or more kinds of glass material with different softening points in the glass/ceramic composite is being tried. Since the composite contracts in accordance with the softening behavior of several kinds of glass, a gentle shrinkage curve can be achieved. This also prevents foaming of the glass, and besides stabilizing the contraction coefficient, it is also an effective method of controlling shrinkage behavior as explained in the next section. In particular, if glass with a high softening point is added, paths for releasing the gasses from the broken down binder can be formed during firing, and it is effective for reducing the residual carbon in the ceramic.

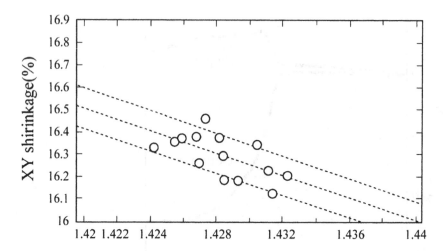

Figure 8-5   The relationship between green sheet density and firing shrinkage rate (x and y axes).

During firing, the reaction between the sample and the setter on which the sample is placed also has a significant impact on controlling the contraction coefficient. During firing, if the substrate sticks to the setter, the firing shrinkage rate varies locally so that contraction does not occur isotropically in the x and y axes and in extreme cases, the board actually bends. To prevent this, it is effective to use a setter with a shape that maintains a small contact area with the substrate. This kind of setter also allows the gas in the atmosphere to flow underneath the substrate offering the additional merit of uniform firing (Fig. 8-6).

Sample (laminated body)

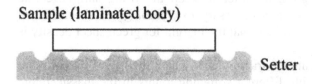

Figure 8-6   Ridged firing setter.

When warping occurs during firing in ceramics that use amorphous glass, it is possible to place a weight on the substrate and to correct the warping by firing it again. However, with ceramics that use crystallized glass it is difficult to reduce warping by weighting the substrate since after firing, there is very little glass residue left in the ceramic so that when it is fired again, the glass does not flow.

(2) Copper/ceramic

In circuit boards, the surface wiring and vias of each layer are copper while the remainder consists of ceramic. When the whole circuit board is viewed as a composite of copper and ceramic, the content of copper in the substrate varies depending on the circuit design of the conductor pattern used, the number of ceramic layers and so on. Figure 8-7 shows the variation in contraction coefficient of the substrate when the copper content changes. In this figure, the contraction coefficient of the ceramic alone is 15.6% while that of the copper alone is 14.8%. If the content of the copper in the substrate exceeds 2 vol%, the contraction coefficient of the substrate approaches that of copper rather than ceramic. Incidentally, the content of copper in the multilayer ceramic circuit boards for computers is around 2 to 3% so that the boards are strongly affected by the contraction coefficient of the copper. The variation in contraction coefficients in substrates that contain copper is caused by a mismatch in the final firing contraction coefficient between the copper and ceramic [2]. The details of contraction mismatch are covered in the next section.

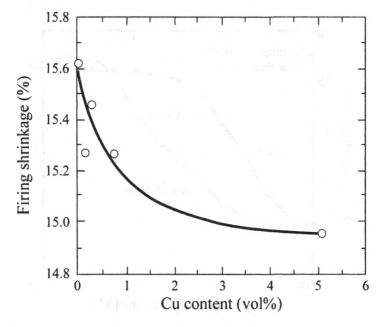

Figure 8-7   The relationship between copper content in the substrate and contraction coefficient.

8.3 Mismatches of firing behavior and firing shrinkage rate

The concept of the matching of firing shrinkage rates of ceramics and conductors was touched upon in Figure 3-1 in Chapter 3. As suggested there, when there is a mismatch in the contraction coefficient between the ceramic and metal materials, defects are formed at their interface [3, 4, 5].

Figure 8-8 shows the results of an investigation into the firing shrinkage rate when laminated bodies of glass/alumina composite and compacts printed with copper paste were fired in a nitrogen atmosphere for 1 hour at temperatures ranging from 500 to 1,050°C at the heating rate: 100°C/min. [6]. The firing shrinkage rates of both the glass/alumina composites and copper paste tend to increase with an increase in firing temperature. However, the amount of shrinkage and shrinkage behavior of both differ. The firing shrinkage rate of the glass/alumina composite increases gradually with the firing temperature, and at around 1,000°C it becomes constant reaching 17.2% at 1,050°C. At 600°C and higher, the copper paste contracts abruptly, and at around 800°C it largely stops contracting. At 1,050°C, the contraction coefficient is 15.9%. Between the two materials, the temperature of the start of firing shrinkage and the final shrinkage coefficient differ. Defining the difference in temperature of the start of firing shrinkage as ΔT, and the difference in final firing shrinkage rate as ΔS, the following chart shows the effect of mismatch between the two.

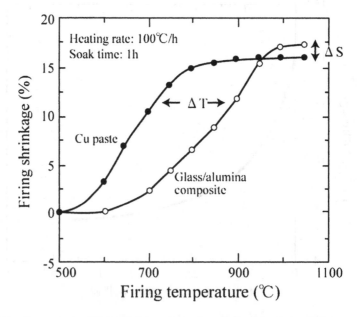

Figure 8-8   Comparison of firing shrinkage behavior of glass/alumina composite and copper paste.

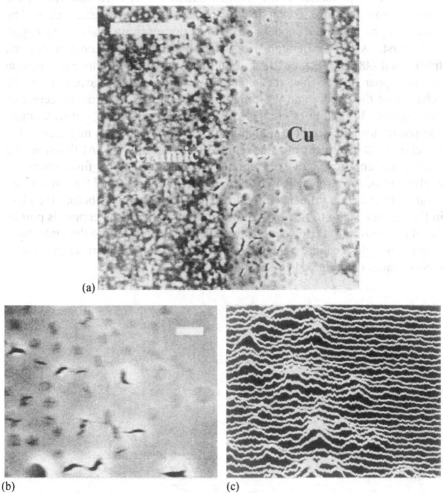

Figure 8-9   A SEM photograph of the copper/ceramic interface of a substrate fired for 1 hour at 1,050°C (Heating rate: 400°C/min.) [Bar = 30 μm] (a), (b) enlarged photograph of the brittle fracture area, and (c) Si distribution using EPMA.

(1) The effect of ΔT

Figure 8-9 (a) is a SEM photograph of the copper/ceramic interface of a substrate fired for 1 hour at 1,050°C at a heating rate faster than 400°C/min. Brittle fractures can be seen in the copper electrode. Figure 8-9 (b) and (c) are enlarged photographs of these brittle fractures and the Si distribution using EPMA. The cracks in the copper and peaks of the silicon match. Since the copper paste does not contain inorganic additives originally, the silicon in the figure can be thought to be the glass phase (SiO_2) in the ceramic. During firing the glass phase in the ceramic migrates to the copper conductor, then due to a mismatch in the thermal expansion of each material in the cooling process, stress is applied to the glass so that cracks occur. The shrinkage process of the copper and ceramic during firing is shown in Figure 8-10. First, as the temperature rises, the copper starts to contract inwards from about 600°C. Next, contraction of the ceramic starts. Since contraction of the copper occurs first, the direction of contraction is governed by the behavior of the copper and the ceramic also contracts towards the center of the copper. In addition, as the firing shrinkage of the glass/alumina composite depends on the liquidation of the glass, the glass migrates in the direction of the copper after the ceramic starts to contract. At this time the contraction coefficient of the ceramic is greater than the final shrinkage coefficient of the copper. It is likely that the morphology of the copper and ceramic interface is influenced by the density of the copper before the glass in the ceramic starts to flow. As shown in Figure 8-9, if the copper is porous the glass seeps into it. However, if the copper densifies before the glass starts to flow, the glass cannot penetrate the copper so that they are segregated at their surfaces (Figure 8-11).

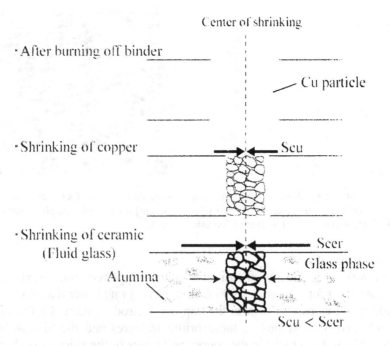

Figure 8-10   The shrinkage process model for copper and ceramic during firing (S_{cer}: ceramic contraction coefficient, S_{cu}: copper contraction coefficient).

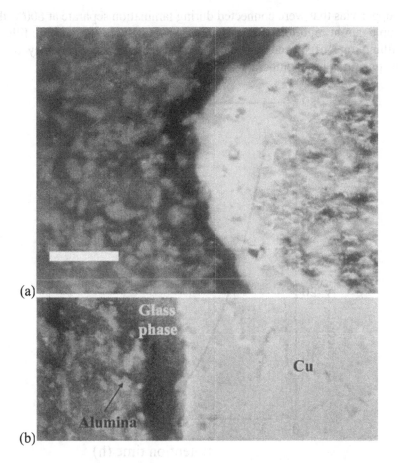

Figure 8-11   SEM photograph of a copper and ceramic interface (example in which the glass is segregated at the surface of the copper) [Bar = 10 μm].

Figure 8-12 shows the electrical resistance of copper and ceramic substrates that are subjected to heat treatment at 800°C with varied times and then finally fired at 1,050°C. When the retention time at 800°C is short, high electrical resistance is seen. When the retention time is short, the copper is porous and is penetrated by the glass, and it is likely that the situation is similar to that in Figure 8-9 causing electrical resistance to rise. Similarly, when the retention time is long, electrical resistance shows a slight tendency to increase, although this is likely to be caused by the formation of closed pores in the copper that occur due to the occurrence of copper vapor described in 8.1.

Figure 8-13 shows the microstructure of the parts around the copper vias of substrates fired at a range of temperatures. Since the copper starts to sinter and contract before the ceramic, when seen at the macro level as in the figure, the copper vias that were connected during lamination separate at 600°C then join up again at high temperatures above 850°C. When the vias are filled, if the filling rate of the copper is low, once the vias separate they do not recover again at high temperature and remain as defects.

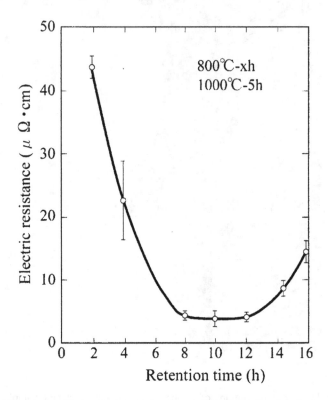

Figure 8-12   The relationship between retention time at 800°C and the electrical resistance of internal copper conductor.

(2) The effect of ΔS
Based on Figure 8-8, at 1,050°C compared with ceramic, copper contracts 1.3% more so that the copper can be expected to undergo compressive stress from the ceramic [7, 8]. Figure 8-14 shows the results of X-ray analysis of copper conductor (220) plane when measured from a variety of incidence angles. If the angle of incidence is big, the angle of diffraction of the copper (220) plane shifts slightly towards a higher angle. Calculating the residual stress from these results shows that compressive stress of around 4 kgf/mm^2 is applied to the copper. As explained in the previous section, ΔS also has an impact on the final shrinkage coefficient of ceramic substrates in which copper is used for the internal wiring.

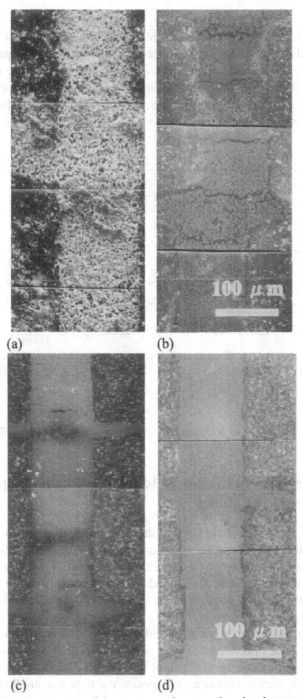

Figure 8-13   The microstructure of the parts around copper vias when heat treated at various temperatures (a) laminated body (b) 600°C, (c) 850°C, and (d) 1,000°C.

In order to reduce the amount of mismatch in the contraction between the two materials, adding alumina or other additives to the copper paste has been suggested. Since alumina powder does not change at the firing temperatures of LTCCs, it can increase the contraction coefficient of copper while it is also effective in strengthening interface adhesion with its anchor effect and by preventing expansion due to foaming.

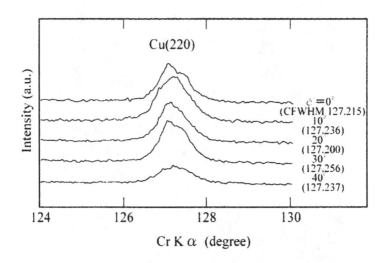

Figure 8-14   The results of X-ray analysis of internal copper conductor (220) plane when measured from a variety of incidence angles.

8.4 Achieving both antioxidation of the copper and elimination of binder

When oxides and metals are fired at the same time, it is normally necessary to control the oxygen partial pressure in the atmosphere. Figure 3-6 in Chapter 3 is a thermodynamic equilibrium diagram of what happens when the oxygen partial pressure is changed. Oxygen partial pressure is controlled using the oxygen concentration in the inert gas and the Thermal Equilibrium of mixed gases.

Pure oxygen has an oxygen partial pressure of 1 atm while that of the atmosphere is 0.2 atm. When oxygen partial pressure is controlled by mixing oxygen in argon or nitrogen, the limit is around 10^{-4} atm. In order to make a lower oxygen partial pressure, it is necessary to use a mixture of H_2/H_2O gases or CO/CO_2 gases.

For example, to achieve an oxygen partial pressure of $10^{-11.7}$ atm at 1,600°C, a mixed gas calculated as follows should be used.

$$CO + \frac{1}{2}O_2 \rightarrow CO_2$$

The Gibbs free energy of reaction in the chemical formula above is indicated by $\Delta G^0 = -282,400 + 86.81T$. If 1,600°C (1, 873 K) is entered, $\Delta G^0{}_{1873K} = -11,9804.9\,(J)$. On the other hand, if the Gibbs free energy of reaction is expressed as an equilibrium constant the result is $\Delta G^0 = -RT\ln K_p$ so that with 1, 873 K, if the calculation result above is substituted, the result is $\ln K_p = 7.7$. If the value K_p and $PO_2 = 10^{-11.7}$ found with the calculation are entered in $K_p = \dfrac{P_{CO_2}}{P_{CO}(P_{O_2})^{1/2}}$ the mixing ratio of the gas $\dfrac{P_{CO_2}}{P_{CO}} = 10^{-2.5}$ can be obtained.

With the same method, using an H_2/H_2O gas mixture, any oxygen partial pressure can be obtained.

In order to fire copper and ceramic at the same time, it is necessary on the one hand to prevent oxidation of the copper which oxidizes readily, and on the other, the organic binder (C) in the ceramic green sheet must be oxidized and driven off. If the copper oxidizes it becomes a nonconductor and does not fulfill the function of a conductor. In addition, if the organic binder is not oxidized and burnt off during firing, and it remains in the ceramic as carbon, the conductivity of the carbon will degrade the insulating property of the ceramic. Figure 8-15 shows the relationship between oxygen concentration in the atmosphere with a temperature in equilibrium, and the chemical reactions of Cu and C, while Figure 8-16 shows the appearance of copper and ceramic substrates fired in a range of atmospheres. When fired in an air atmosphere or similar in Range I, the binder in the ceramic is eliminated, however since the copper oxidizes, the wiring turns black due to the formation of CuO. In Range II with a lower oxygen concentration than Range I, the ceramic binder is eliminated and appears white, while the wiring appears reddish brown since Cu_2O is formed. In Range IV, an atmosphere of inert gas with high purity or similar, copper oxidation is suppressed and low wiring resistance can be achieved, although the substrate turns black since the organic binder in the ceramic part is not broken down and it carbonizes.

When the atmosphere is adjusted to an intermediate range (Range III), elimination of the carbon in the ceramic and prevention of oxidation of the copper can both be achieved [9].

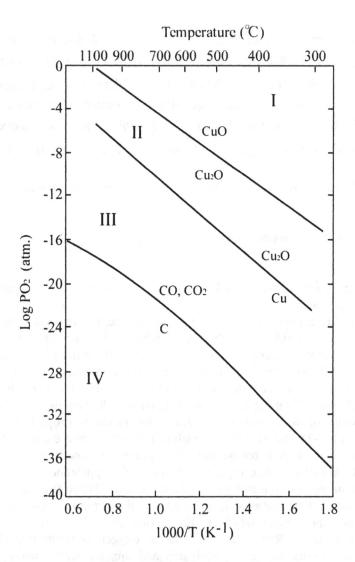

Figure 8-15   The relationship between oxygen concentration in the atmosphere with a temperature in equilibrium, and the chemical reactions of Cu and C.

The four combinations above can be observed in an equilibrium, however in practice the ceramic part may appear black and the conductor part may appear reddish brown. If a glass/alumina composite using glass with an extremely low softening point is fired in Range II, the copper turns reddish brown and in the ceramic part, the carbon that might be expected to break down and be driven off, is trapped by the softened glass and turns black. Figure 8-17 shows an XMA profile of a blackened glass/alumina composite. The peaks for Si and C detected nearly match, and it can be confirmed that the glass traps the carbon [10].

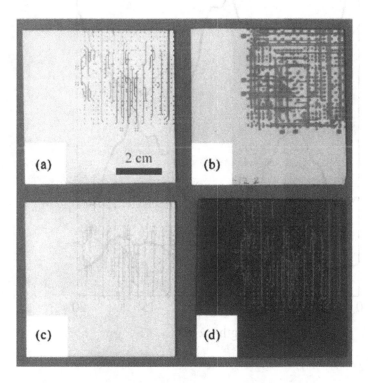

Figure 8-16   External appearance of copper/ceramic substrates fired in various atmospheres. Firing conditions; (a) Range I, (b) Range II, (c) Range III, and (d) Range IV in Figure 8-15.

In order to eliminate carbon effectively, it is necessary to control the firing environment and to form a path for the carbon to escape to the outside after it oxidizes and breaks down. If the path of the carbon is blocked, the substrate blackens. Figure 8-18 shows the results of cross-sectional observation of a cut copper/ceramic substrate fired at 700°C, and element distribution of Cu and C using XMA[11].

As can be seen from the figure, carbon is concentrated in the area of the interface between the copper and ceramic. This is probably due to the fact that since copper begins sintering faster and reaches its end temperature earlier than ceramic, the sintered copper on top of the unsintered ceramic blocks the openings that release the gas. In order to eliminate the carbon under the copper conductor completely, it is necessary to break down the binder sufficiently at about 400°C before sintering of the copper occurs.

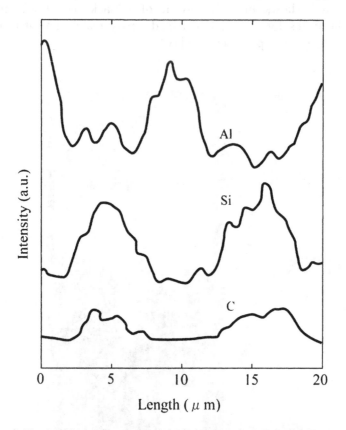

Figure 8-17   A XMA profile of a blackened glass/alumina composite.

(a)

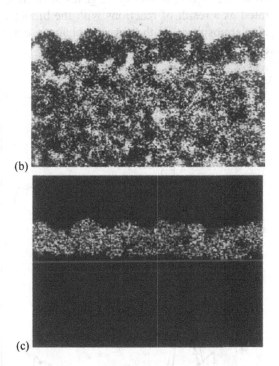

(b)

(c)

Figure 8-18   The results of cross-sectional observation of a cut copper/ceramic substrate fired at 700°C, (a) SEM photograph, (b) C distribution, and (c) Cu distribution.

As described above, controlling the gas composition of the firing environment is important for preventing oxidation of the copper and eliminating the binder at the same time, although it is necessary to bear in mind that the composition of the gas introduced and that of the gas in the furnace differ. The following are three causes of changes in the composition of gas in the furnace.

- Reaction with the ceramic substrate
- Flow of gas in the electric furnace
- Impurities inside the electric furnace and on its surfaces

When fired, the organic binder in the ceramic undergoes thermal decomposition and at the same time the atmospheric gas that adjusted the oxygen partial pressure reacts with the material creating new gas in the furnace. Figure 8-19 shows the results of gas chromatography of the gases generated during firing in the furnace. At the time of monitoring, heat treatment was maintained chiefly at around 400°C where thermal decomposition of the binder becomes active, and around 800°C where reactions between the atmospheric gas and impurities in the substrate occur readily. As shown in the figure, a variety of gases such as CO, CO_2, CH_4, and H_2 are generated as a result of reactions with the binder in the ceramic. In other words, the atmosphere in the furnace changes constantly when a ceramic substrate is introduced into the furnace.

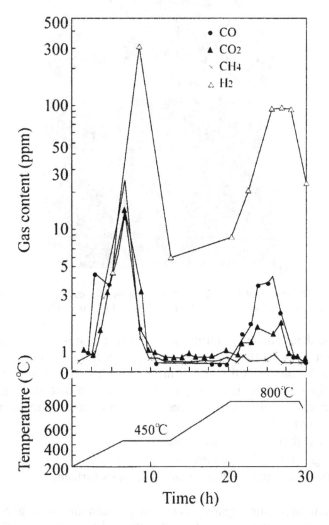

Figure 8-19   The measuring result of gases arising from a copper/ceramic substrate.

For the same reason, the placement of samples in the electric furnace requires careful consideration. Figure 8-20 shows the flow of gas in a continuous furnace with a conveyor of the type frequently used for firing LTCCs. In a continuous furnace with a conveyor there are some ten heater zones and by changing the temperature setting of each zone, a temperature profile for firing can be made. The samples are then passed through the furnace on the conveyor to be fired. To adjust the atmosphere in the electric furnace, nitrogen gas is supplied continuously in the furnace. In addition, to shut off the outside from the atmosphere, a gas curtain of large volumes of nitrogen gas is pumped at the entrance and exit of the furnace. There are also exhaust vents at the top of the furnace for venting the contaminated reactant gases generated by the samples. As shown in the figure, the contaminated gas generated by the samples that are conveyed in first forms a flow in the electric furnace that flows over the samples behind so that although the first samples are fired properly, the subsequent samples are contaminated and become blackened. It is necessary to form a flow in the furnace such that gases including contaminants are vented appropriately.

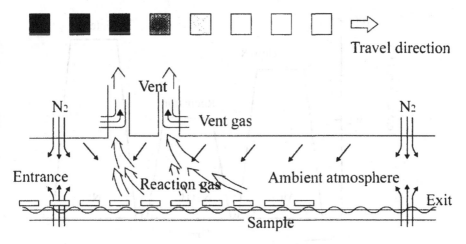

Figure 8-20   Flow of gas in a continuous furnace with a conveyor.

The third cause of changes in the composition of gas in the furnace is the impurities in the electric furnace and on its surfaces. It is common in the field of ceramics for firing to be poor in recently purchased new electric furnaces. When atmospheric furnaces are first purchased, the moisture content in the refractory material in the electric furnace is not sufficiently eliminated, so that at high temperature, a water vapor atmosphere may be formed. In addition, when firing ceramic with high vapor pressure such as lead monoxide, when the furnace has been used sufficiently long for accretions of lead oxide to accumulate in the furnace, high density lead monoxide vapor forms naturally so that an appropriate firing environment can be obtained. Normally nitrogen furnaces are used for LTCCs, however the atmosphere is not stable when the electric furnace is new. If an electric furnace is used for a certain period of time, the walls of the furnace become thickly covered in organic resin that has not burnt completely. This carbon impurity reacts with the oxygen contained as a contaminant in the gas introduced into the furnace, and this effectively acts to maintain a low oxygen concentration at all times.

The composition of gases in the ceramic substrate also differs locally. For example, the binder resin in the copper paste reacts with the atmosphere to form a new gas composition so that a different atmosphere is created just in the vicinity of the copper parts.

As described above, controlling the gas composition and making an atmosphere in which the copper does not oxidize and the carbon ingredients of the binder resin and so on oxidize is a method of both preventing oxidation of the copper and eliminating the binder.

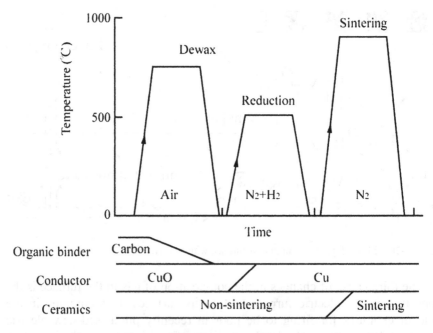

Figure 8-21   Conceptual diagram of the process for cofiring copper and ceramic substrate.

In addition to this method, the following technique can be applied with separate processes using different heat treatments suited to each material (refer to Figure 8-21)[12]. As shown in the figure, the laminated body is heat treated in the atmosphere at 600°C as the first step to burn off and completely eliminate the organic binder resin in the ceramic. At this time, the copper is transformed to copper oxide (CuO). Next, the substrate undergoes heat treatment in hydrogen gas (or nitrogen gas containing hydrogen) at 500°C. With this process, the copper oxide (CuO) is completely deoxidized into Cu. Finally with firing at 1,000°C in nitrogen, the ceramic and copper are both densified. Since delicate control of the gas composition in the atmosphere is not required with this method, preventing oxidation of the copper and achieving elimination of the binder can both be achieved easily at the same time. However, in this process the Cu is transformed into CuO in the dewaxing process at 600°C in the air atmosphere, and in the next firing process it is transformed into Cu again. When the Cu oxidizes it undergoes volume expansion and when it deoxidize it shrinks so that stress is generated between the ceramic and copper. This can cause problems with delamination.

8.5 Non-shrinkage process

After firing of the ceramic substrate, thin film wiring is formed on it and various kinds of components are mounted on it. If there is variation in the contraction coefficient of the surfaces in the ceramic, the thick film conductor pattern and thin film wiring artwork formed on the substrate of those parts will not match up, and lack of conduction will be a problem. In addition, if shrinkage occurs isotropically in the x and y axes and the contraction coefficient differs from the circuit design dimensions, when the origin of coordinates is provided on one edge of the substrate and there is significant misalignment on opposite edges of the substrates, it may be difficult to mount components at the edges of the substrate. For this reason, it is necessary to fire the substrates with the same isotropic contraction coefficient each time.

The "non-shrinkage process" is a method in which firing shrinkage in the x and y axes is suppressed, allowing shrinkage only in the z axis. Since shrinkage does not occur across the surface, the conductor pattern after firing retains the same dimensions as when printed, and the misalignment described above does not arise. The following two methods are commonly known [13, 14].

- Firing with a weight added to the z axis
- Sandwiching the laminated body between material such as alumina that does not sinter at LTCC firing temperatures, to constrain movement in the x and y axes for firing

Incidentally, the actual contraction coefficient is 0.1% in the x and y axes and 40 - 50% in the z axis.

8.6   Cofiring process and future LTCCs

The cofiring process is a characteristic process in which materials suited to LTCCs are selected and this can be thought of as the key to deciding the product form of the LTCC. Here we focus on the cofiring process and the developmental direction of future LTCCs. The overall prospects for the future of LTCCs are covered in Chapter 10.

This chapter covered only the cofiring technology for alumina and glass composites and copper conductor, although as already indicated in Chapter 1, in future LTCCs, it will be necessary to cofire a variety of dielectrics with different dielectric constants and resistor materials. In order to achieve this, it will first be necessary to be able to sinter each separate material carefully using the same firing conditions to achieve the required characteristics (dielectric characteristics and resistance characteristics). Controlling shrinkage mismatch between dissimilar materials and interface reactions is the next problem. With LTCCs, firing conditions (temperature and atmosphere) are regulated by the melting point of the conductive material and resistance to oxidation.

The firing temperature of copper conductor is about 1,000°C and it must be fired in a reducing atmosphere. Since alumina/glass composite ceramic is very resistant to reducing, no difference in its insulating properties is seen when fired in the air atmosphere or in a reducing atmosphere, but high dielectric constant material in the barium titanate series deoxidizes easily becoming a semiconductor in a reducing atmosphere. Lead series relaxers show a high dielectric constant as well as having a firing temperature of about 900°C so they are promising as dissimilar materials for cofiring although they transform readily to metallic lead in a reducing atmosphere. As explained in Chapter 4 concerning resistor materials, since conductive oxides deoxidize readily (for example, RuO_2 deoxidizes to Ru), controlling their value of resistance is difficult.

On the other hand, if silver, silver/palladium, or gold are used as conductive material, it is not necessary to take deoxidation of the material into account since they can be fired in the air atmosphere, and a wide choice of materials is possible. Since it is not necessary to bear in mind driving off the binder as with copper materials explained above, the manufacturing process is simple. Silver and palladium are not entirely without problems, such as silver oxide due to low temperature oxidation diffusing into the glass, and dilation through oxidation of the palladium [15]. However, compared with copper, their problems are really insignificant.

As explained above, in LTCCs that incorporate a variety of materials with embedded passive components, especially where fine wiring is used, it will be beneficial to develop LTCCs that use mainly silver. However, in circuit board applications consisting of conductor and insulating materials with fine wiring, low resistance, highly migration resistant copper conductors will also likely be beneficial. Depending on their application, various materials will continue to be used side by side.

References

[1] K. Niwa, N. Kamehara, H. Yokoyama, and K. Kurihara, "Multilayer Ceramic Circuit Board with Copper Conductor," Advances in Ceramics, Vol. 19, (1986), pp. 41-48.

[2] Y. Imanaka, "Multilayer Ceramic Substrate, Subject and Solution of Manufacturing Process of Ceramics for Microwave Electronic Component, Technical Information Institute, (2002), pp. 235-249.

[3] R. R. Tummala, "Ceramic and Glass-Ceramic Packaging in the 1990s," J. Am. Ceram. Soc., Vol. 74, (1991), pp. 895-908.

[4] G. –Q. Lu, R. C. Sutterlin, and T. K. Gupta, "Effect of Mismatched Sintering Kinetics on Camber in Low Temperature Cofired Ceramic Package," J. Am. Ceram. Soc., Vol. 76, No. 8 (1993), pp. 1907-14.

[5] T. Cheng and R. Raj, "Flaw Generation During Constrained Sintering of Metal-Ceramic and Metal-Glass Multilayer Films," J. Am. Ceram. Soc., Vol. 72, (1989), pp. 1649-55.

[6] Y. Imanaka and N. Kamehara, "Influence of Shrinkage Mismatch between Copper and Ceramics on Dimensional Control of Multilayer Ceramic Circuit Board," J. Ceram. Soc. Jpn., Vol. 100, No. 4, (1992), pp. 560-564.

[7] C. H. Hsueh and A. G. Evans, "Residual Stress and Cracking in Metal/Ceramic Systems for Microelectronics Packaging," J. Am. Ceram. Soc., Vol. 68, No. 3, (1985), pp. 120-27.

[8] R. K. Bordia and R. Raj, "Sintering Behavior of Ceramic Films Constrained by a Rigid Substrate," J. Am. Ceram. Soc., Vol. 68, No. 6, (1985), pp. 287-92.

[9] Y. Imanaka, A. Tanaka, and K. Yamanaka, "Multilayer Ceramic Circuit Board –Wiring Material: From Copper To Superconductor-," FUJITSU, Vol. 39, No. 3, June, (1988), pp. 137-143.

[10] N. Kamehara, Y. Imanaka, and K. Niwa, "Multilayer Ceramic Circuit Board with Copper Conductor", Denshi Tokyo, No. 26, (1987), pp. 143-148.

[11] K. Niwa, N. Kamehara, K. Yokouchi, and Y. Imanaka, "Multilayer Ceramic Circuit Board with a Copper Conductor," Advanced Ceramic Materials, Vol. 2, No. 4, Oct. (1987) pp. 832-835.

[12] T. Ishida, S. Nakatani, T. Nishimura, and S. Yuhaku, "A New Processing Technique for Multilayered Ceramic Substrates with Copper Conductors," Advances in Ceramics, Vol. 26, (1987), pp. 467-480.

[13] H. T. Sawhill, R. H. Jensen, K. R. Mikeska, "Dimensional Control in Low Temperature Co-fired Ceramic Multilayers," Ceramic Transactions, Vol. 15, (1990), pp. 611-628.

[14] K. R. Mikeska, and R. C. Mason, "Pressure Assisted Sintering of Multilayer Packages," Ceramic Transactions, Vol. 15, (1990), pp. 629-650.

[15] Y. Imanaka, "Material Technology of LTCC for High Frequency Application" Material Integration, Vol. 15, No. 12, (2002), pp. 44-48.

Chapter 9
Reliability

After cofiring is finished, first the substrate is visually inspected and its firing shrinkage rate is evaluated. In the visual inspection, items related to physical defects such as delamination, warping, cracks and the like are checked, and items related to the external color of the ceramic parts that can be observed at the stereoscopic microscope level such as coloration, oxidation of the metal and so on are evaluated. To check the contraction coefficient, the dimensions in the x and y axes and z axis are measured. The substrate is examined for deviation from the design, and is checked to see whether it is in the permissible range. Next, simple checks of the electrical properties of the substrate are carried out, such as its conductive properties and conductor resistance, the interlayer insulation resistance between the ground conductors, characteristic impedance and so on. Additionally, the quality of the substrate is checked from the material point of view, for density, strength, microstructure and the like. When these quality checks are finished, reliability testing of the fired substrate is carried out, and only substrates that clear these tests are approved as products. The following test conditions are typical of those generally used for reliability testing.

- Temperature cycle test (gas):
    -60°C, 20 min. ↔ 150°C, 20 min., 1,000 cycles
- Thermal shock test (liquid):
    -60°C, 20 min. ↔ 150°C, 20 min., 1,000 cycles
- Pressure cooker test:
    110°C, 85% RH, 1.2 atm., 500 h
- High temperature and high humidity bias test:
    85°C, 85% RH, DC 5 V, 2,000 h
- Heat exposure test: 150°C, 2,000 h

The majority of defects found in reliability and quality testing arise at the interface between the copper and ceramic. Since more than 95% of the circuit board overall is ceramic, there is a tendency to evaluate only the reliability of the ceramic part. However, LTCCs are ceramics formed with fine three dimensional conductor wiring (a composite material of metal and ceramic (see Fig. 9-1)), and it is difficult to explain the various characteristics and reliability issues with the qualities of the ceramic alone.

Since the thermal characteristics and mechanical properties of the conductor differs significantly from those of the ceramic, the conductor has a significant impact on the macro and micro characteristics of the entire LTCC. This chapter provides an example of the impact on the reliability of the LTCC overall when there is a mismatch between the thermal and mechanical properties of the ceramic and conductor, especially when copper is used.

The details of the mechanical and thermal characteristics of the ceramic alone are covered in Chapter 2. The ceramic used in LTCCs is often a composite of glass and ceramic, and the composite mixing rules for obtaining the various characteristics covered in Chapter 2 can be applied to the discussion in this chapter. However, although ceramic has an almost isotropic composite structure, with circuit boards (conductor and ceramic composite), there is no order in the arrangement of the conductor parts and the fact that they are present as a continuum in the ceramic is the point of difference.

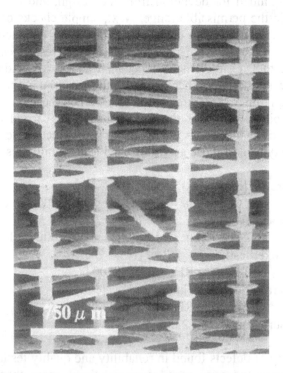

Figure 9-1   Three dimensional copper wiring of multilayer ceramic circuit board for main-frame computer (part of the ceramic was etched away with HF).

9.1 Thermal shock of LTCCs

If heat treatment such as rapid cooling or rapid heating is carried out with even a single material, thermal stress due to the temperature difference between the surface and the inside of the material arises, and thermal shock occurs. In addition to the temperature difference in these locations, since LTCCs are composites of conductor metals and ceramic, warping arises due to the difference in elongation arising from the difference in thermal expansion of each material, and stress arises between both materials. When heated rapidly, the compressive stress on parts with high thermal expansion generates tensile stress on parts with low thermal expansion.

The maximum stress σ arising at the surface of each material is expressed with the following equation [1, 2]. Table 9-1 shows the different physical properties of LTCC related materials.

$$\sigma = \frac{E\alpha\Delta T}{1-v}$$

E: Young's modulus, α: thermal expansion, ΔT: temperature difference, and v: Poisson's ratio

Table 9-1   The physical properties of materials related to LTCCs that are concerned with thermal stress.

	Young's modulus E (GPa)	Thermal expansion coefficient α ($\times 10^{-6}$/°C)	Poisson's ratio v
Ag	75	19.7	0.38
Cu	120	16.5	0.37
W	360	4.5	0.35
Alumina	380	8.0	0.24
Pyrex glass	70	3.0	0.16
Glass/alumina composite	90	4.0	0.17

Figure 9-2 shows the copper/ceramic interface after applying thermal shock of ΔT 250°C to an LTCC with copper conductor (a), and the copper/ceramic interface after conducting a bending strength test on a sample that has undergone thermal shock (b). Immediately after undergoing thermal shock, cracks and so on cannot be observed, and no changes can be seen in the microstructure of the material. On the other hand, in the sample that has undergone the bending strength test, many cracks and flaws can be seen in the ceramic part. However, none can be seen in very close proximity to the copper/ceramic interface. This speaks of the significant bonding strength of the copper and ceramic. Incidentally, while the bending strength of the ceramic part alone is 150 to 200 MPa, the bending strength of LTCC with copper conductor is 80 MPa.

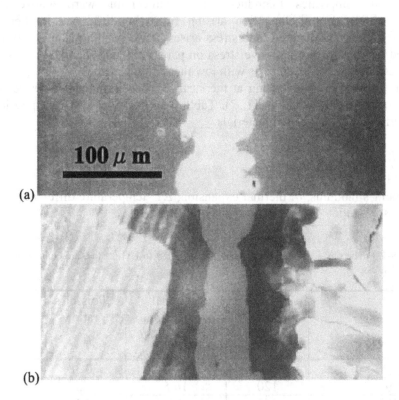

Figure 9-2   The copper/ceramic interface after applying thermal shock of ΔT 250°C to an LTCC with copper conductor (a) and the copper/ceramic interface after conducting a bending strength test on a sample that has undergone thermal shock (b).

9.2 Thermal expansion and residual stress of LTCCs

Figure 9-3 shows a cross section of the surface via of a copper/ceramic substrate at room temperature and after heating at 200°C. At room temperature, the via is at about the same height as the surrounding ceramic. If the substrate rises in temperature to 200°C, the via is raised up by about 0.4 μm.

When a composite made of two materials such as LTCC is heated, if both materials are in a free state, elongation of each material occurs to the extent of $\alpha\Delta T$ of each. However, in actual LTCCs, since the conductor metal and ceramic are mutually constraining, internal stress occurs due to the strain between both materials (refer to Figure 9-4). For example, if the thermal expansion coefficient of the metal/ceramic composite is α_{co}, the thermal expansion coefficient of the ceramic is α_{ce}, and temperature change ΔT is used, and strain of the ceramic is ε_{ce}, the balance of strain is expressed with the equation $\alpha_{co}\,\Delta T = \alpha_{ce}\Delta T + \varepsilon_{ce}$

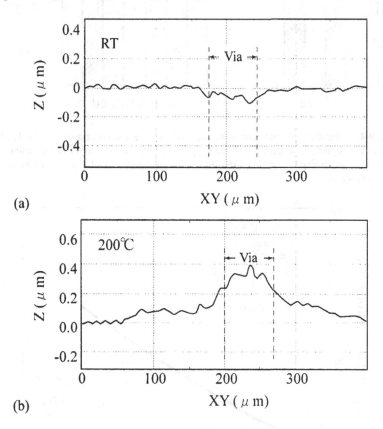

(a)

(b)

Figure 9-3   The surface profile of a via in a copper/ceramic substrate (before and after thermal deformation, RT-200°C).

In the same way, if we find the balance of strain of the conductor metal and take the difference between both equations, the strain between both materials is expressed as the difference between the thermal expansion coefficients in the equation $(\alpha_{ce} - \alpha_m)\Delta T = \varepsilon_m - \varepsilon_{ce}$

Figure 9-5 shows the thermal expansion behavior of a copper via in an LTCC. Since we are measuring the expansion of the top of the copper via, the ε_{ce} part of the equation above is not taken into account. However, the thermal expansion coefficient of copper via conductor when actually measured is $12 \times 10^{-6}/°C$, and the difference between the thermal expansion of copper alone ($16.5 \times 10^{-6}/°C$) and the thermal expansion of ceramic ($4.0 \times 10^{-6}/°C$) is shown with a nearly equivalent value.

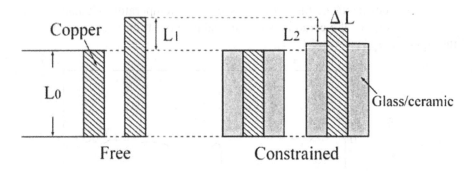

Figure 9-4   The thermal deformation of materials in free and constrained states (L_0: initial length of the copper, L_1: expansion of the copper in a free state, L_2: relative expansion of the copper in a constrained state, and ΔL: difference in expansion of the copper in free and constrained states.

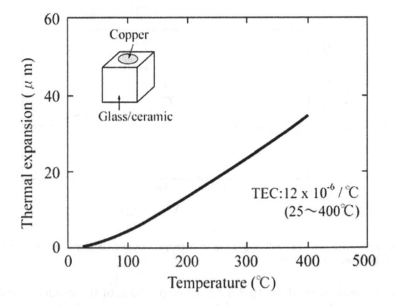

Figure 9-5   The thermal expansion behavior of copper via parts.

As noted above, the strain that occurs is determined by the difference in thermal expansion of the materials. As indicated in Figure 9-6, since LTCCs are combinations of glass/ceramic composites and metals such as gold, silver, copper and so on, the mismatch in thermal expansion is greater than that of the combination of materials of HTCCs (insulator; alumina, conductor: Mo or W). However, the Young's modulus of the material used in HTCCs, Mo or W, is about three times greater than that of the metals used in LTCCs. In other words, in LTCCs, the amount of strain of the conductor part is greater, although the stress arising from the strain is less. The results of stress simulation make it clear that the total internal stress arising is less in LTCCs compared with HTCCs, and reliability concerning thermal characteristics it higher [3]. However, when seen from a micro point of view, protrusion of the via part occurs at high temperatures as shown in Figure 9-3. This can cause problems with adherence when forming thin film conductors or the like on the surface of the LTCC (refer to Figure 9-7). It is important to suppress movement of the via conductor alone by strengthening the adherence between the via conductor and ceramic.

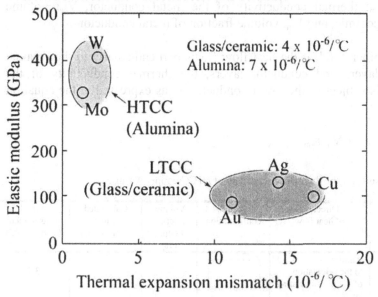

Figure 9-6   The mismatch between ceramic and conductor expansion and the relationship of the Young's modulus and conductive materials of HTCCs and LTCCs.

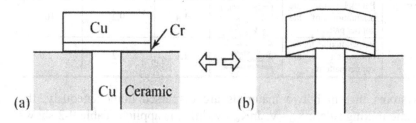

Figure 9-7   Example of a defect affecting the thin film wiring due to protrusion of the copper via part (a) at room temperature, and (b) at high temperature.

9. 3 Thermal conductivity of LTCCs

In composites of metal and ceramic such as LTCCs, thermal conductivity depends significantly on the thermal conductivity, volume fraction, and composite form of the constituent materials [4, 5]. If the via conductors formed between the insulating layers are excluded, multilayer circuit boards exhibit a layered structure of ceramic insulating layers and conductor layers. If the ceramic insulating layers and conductor layers are arranged parallel to the direction of heat flow, the heat flow largely flows through the high thermal conduction layers, expressed with the following equation.

$$K_{co} = V_{ce}K_{ce} + V_mK_m$$

K_{co}: Thermal conductivity of the composite, K_{ce}: thermal conductivity of the ceramic, K_m: thermal conductivity of the metal conductor, V_{ce}: volume fraction of ceramic, and V_m: volume fraction of metal conductor

On the other hand, if the heat flow flows perpendicularly to the ceramic insulating layers and conductor layers, the thermal conductivity of the composite is subject to the lower conductivity, as expressed in the equation below.

$$1/K_{co} = V_{ce}/K_{ce} + V_m/K_m$$

Table 9-2 Results for the thermal conductivity of LTCC copper wiring

Material	Direction of heat flow	Sample configuration	Volume fraction of copper (%)	Calculated value (J/m · s · °C)	Actual measurement (J/cm · s · °C)
Glass/ceramic composite	Parallel to the orientation of the insulating layer		-	-	3.7
	Perpendicular to the orientation of the insulating layer		-	-	3.2
Copper wiring LTCC	Parallel to the orientation of the copper pattern		1.4	9.2	10.5
	Perpendicular to the orientation of the copper pattern		1.1	3.2	3.0

In systems in which two materials are dispersed homogeneously, the logarithmic mixing rule $\ln K_{co} = V_{ce}\ln K_{ce} + V_m\ln K_m$ is applied. Table 9-2 shows the results of measurement of the thermal conductivity of LTCC copper wiring under a variety of conditions. The actual measurement closely matches the calculated values found using the composite rule.

Therefore, when considering thermal design for high reliability LTCCs, it is necessary to bear in mind the arrangement and orientation of the conductor wiring. In addition, if the metal is arranged in parallel with the heat flow, it is possible to form structures known as thermal vias to exploit the higher thermal conductivity, allowing the heat generated to escape efficiently around the periphery of the LTCC.

References

[1] B. A. Boley and J. H. Weiner, Theory of Thermal Stresses, Jhon Wiley & Sons, Inc., New York, (1960).

[2] S. Timoshenko and J. N. Goodier, Theory of Elasticity, 2nd ed.; Chapter 14, MacGrraw-Hill Book Co. , New York, (1951).

[3] Y. Imanaka, K. Hashimoto, W. Yamagishi, H. Suzuki, N. Kamehara, K. Niwa, and K. Kurosawa, "Reliability of Copper Vias for Multilayer Glass/Ceramic Circuit Boards," Proc. Electric Ceramic Conference, Shonan Institute Technology (1991).

[4] M. Kinoshita, R. Terai, and H. Haidai, "Thermal Conductivity of Glass Copper-Composite," J. Ceram. Soc. Jpn., Vol. 88, No. 1, (1980), pp. 36-49.

[5] W. D. Kingery, H. K. Bowen and D. R. Uhlmann: Introduction to Ceramics 2nd ed. (John Wiley & Sons, Inc., (1976), pp. 634.

Chapter 10

Future of LTCCs

10.1 Introduction

With the explosive growth of mobile phones, related functions and services are expanding rapidly, and recently it has become possible to send and receive even large images at high speeds. The advances being made in wireless communication technologies require close attention. Furthermore, in order to develop a ubiquitous computing network of the future, progress is underway towards realizing a communications service environment in which networks, terminal devices, and contents can be used freely and reliably anywhere without any sort of restriction [1]. In this society of ubiquitous networks, high frequency, high speed wireless transmission technologies and miniaturization of multifunction terminal devices will likely be key technologies for use as communication tools and in systems development.

Small circuit modules with excellent high frequency characteristics, embedded and integrated with several passive components, are one of the technologies in the embedded hardware category now being developed [2, 3]. These are small, monolithic modules with circuit elements such as capacitors, resistors, inductors and so on formed in three dimensional multilayer structures and with transmission lines, decoupling capacitors, filters, baluns, and other integrated components with high frequency circuit functions. For passive components like capacitors and filters, it is desirable to use a ceramic that has optimal dielectric characteristics for the function required. Compared with other materials, the high frequency loss of ceramics (LTCCs) is small and since they can be combined easily with dissimilar materials, they are the most suitable candidate as a constituent material for hardware used in communication networks [4, 5].

In order to achieve higher frequencies, higher integration and multifunctionality with smaller size for hardware in the future, it is necessary to devise further improvements to the characteristics of LTCCs themselves to suit them to future applications.

The next section looks at the technology development of LTCCs that must be carried out in order to meet the needs of future ubiquitous network applications.

10.2 Technology development of LTCCs for the future

In order to achieve higher frequencies and smaller, multifunctional integrated passive components, and circuit boards with embedded passive components, it is necessary to proceed with materials technology development and process technology development from the perspectives described here. In applications for high frequency ranges above the microwave band, the homogeneity of the dielectric material, the surface profile of the conductor [6], and the dielectric/conductor interface configuration can be considered relevant to avoiding localized concentration of electric fields within the device, so that it will be necessary to carry out technology development combined with electromagnetic field simulations at high frequency.

Technology development for materials and processes must also be low cost for reasons of manufacturing.

10.2.1 Materials technology development

(1) Conductive material
To reduce conductor loss in high frequency ranges, it is necessary to take an approach that reduces conductor resistance to the minimum (refer to Chapter 1). Since the inductance of the conductor inside increases at high frequencies, current flows only near the surface of the conductor layer. The thickness of the area where the current flows is called skin depth. Figure 10-1 shows the relationship between the frequency of each type of conductor and the skin depth. The relationship with skin depth (δ_m) is in accordance with the formula below, and there is a tendency for the skin depth to become shallower as the frequency increases with materials that are not magnetized.

$$\delta_m = \sqrt{\frac{\rho}{\pi \mu_r \mu_0 f}}$$

ρ: conductor specific resistance (x $10^{-8}\Omega \cdot$ cm), μ_0: permeability of free space ($4\pi \times 10^7$ H/m), μ_r: permeability (1 for non-magnetic material), f: frequency

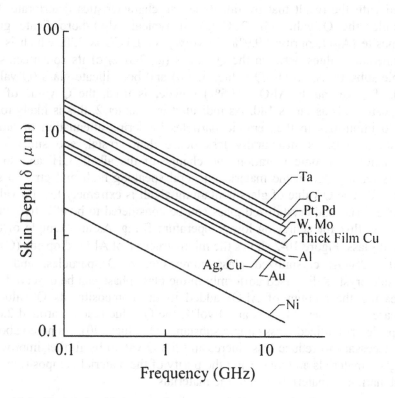

Figure 10-1   The relationship between frequency and skin depth of various conductors.

Since LTCC conductors (especially internal conductors) are produced from conductive paste using thick film processing, defects such as voids, pores and the like form readily within the conductor. Due to the effects of inorganic material ingredients previously added to the conductive paste, the measured resistance value will be bigger compared to the original material property value. Furthermore, since the surface roughness of the conductor is comparatively greater with thick film processing, the conduction path of the conductor (the length of the surface part) becomes longer, and locations where unwanted field concentration occur increase so that transmission loss becomes greater. For this reason, it is important to keep defects inherent in the conductive material to the absolute minimum and to achieve resistance close to that of the bulk material. Since low resistance in the conductor itself is most effective in reducing loss, it is also effective to use high-temperature superconducting material for the wiring.

(2) Dielectric material

For dielectrics as well, materials development aimed at reducing dielectric loss will be necessary. To allow low temperature firing of LTCCs, glass is added with the result that overall dielectric characteristics deteriorate. For example, the Q value (@ 2 GHz) of typical Al_2O_3/borosilicate glass composite (Al_2O_3 purity: 99.9%, 20 vol%) for LTCC is 320, which is the intermediate values between the Q values (@ 2GHz) of its constituents as simple substances, Al_2O_3 (Q value: 3,000) and borosilicate glass (Q value: 150). If lower purity Al_2O_3 (97.5%) powder is used, the Q value of the composite falls as far as 180. As indicated in Chapter 2, this is likely to be due to impurities in the ceramic particles hindering attenuation of lattice vibration. Glass is structurally less dense than crystal, and since ionic conduction and ionic vibration and change of the alkali, OH⁻ and so on occurs readily within the material, general speaking high dielectric loss is indicated. The Q value of glass containing alkali is extremely low (Q value: 95 @ 2 GHz) and crystallized glass can be considered to be effective when added as flux for achieving low temperature firing. As an example of this type of glass, Figure 10-2 shows the microstructure of Al_2O_3/diopside (CaO · MgO · $2SiO_2$) crystallized glass composite. Al_2O_3 particles, and fine diopside crystals dispersed uniformly in the glass phase can be observed. By increasing the amount of Al_2O_3 added to the composite, its Q value is increased and when Al_2O_3 is at 40 vol%, the Q value reaches around 2,000 (diopside crystallized glass simple substance Q value: 340). As noted above, it is necessary to reduce loss (increasing the Q value) by further improving the glass materials and increasing the purity of the material composition, the combination of materials, and of the materials.

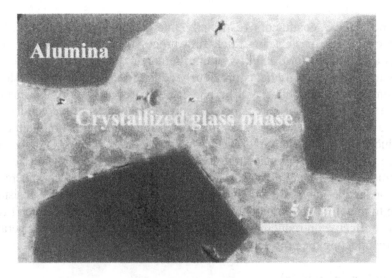

Figure 10-2   The microstructure of Al_2O_3/diopside (CaO · MgO · $2SiO_2$) crystallized glass composite.

In order to provide a single module with a variety of high frequency functions, materials with low dielectric loss and varying dielectric constants are necessary. Since LTCCs were originally developed for use as circuit boards in transmission lines, the dielectric constant of their constituent materials is 10 or less. By making the wavelength of high frequency electromagnetic waves inversely proportional to the square root of the dielectric constant of the transmission medium, it is possible to achieve a smaller component, if for example, the material used for a filter has a high dielectric constant. Up till now, microwave dielectrics have been used for filters, such as $Ba(Mg_{1/3}Ta_{2/3})O_3$, $Ba(Zn_{1/3}Ta_{2/3})O_3$, or $BaO\text{-}TiO_2$, fabricated with a $TE_{01\delta}$ resonator structure. These dielectrics have a high dielectric constant and high Q value, but as their firing temperature is also high at around 1,500°C, when fabricating multilayered structures such as LTCCs that are good for integration, metals with low electrical resistance such as copper and silver cannot be used as conductors (required firing temperature: around 1,000°C). In other words, for integration in multilayered structures, it is necessary to develop LTCC dielectric ceramic materials for filters that simultaneously meet all the requirements for a low firing temperature, low dielectric loss (high Q value), and high dielectric constant. Figure 10-3 shows a graph of the dielectric constant, dielectric loss, and firing process temperature of various kinds of ceramic. Microwave dielectrics such as $Ba(Mg_{1/3}Ta_{2/3})O_3(BMT)$, $BaTi_4O_9$, and $ZrSnTiO_4$ show a high dielectric constant and high Q value, however they do not meet the requirements for processing temperature. The alumina and SiO_2 used as a high frequency insulating material have a high Q value, however they have a low dielectric constant and since they have a high processing temperature, they also do not meet the requirements. Glass and the current LTCCs clear the processing temperature requirements although their dielectric constant and Q value is low. Therefore the emergence of new materials that meet the three requirements noted above is desirable.

Materials with a high dielectric constant are also required for decoupling capacitors and the antenna layers, and it is hoped that LTCCs will be developed to meet these various requirements. It is important to increase the number of varieties of LTCC materials available for different applications.

The high level of integration of active devices is impressive, and the heat release of each device reaches several dozen W so that when LTCCs are used as substrates for mounting devices, it is necessary to combine materials with excellent heat dissipation [7].

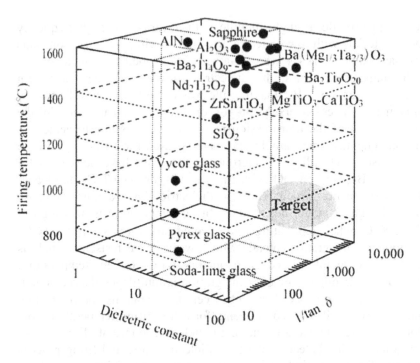

Figure 10-3   The dielectric constant, dielectric loss, and firing process temperature of various kinds of ceramic.

10.2.2 Process technologies

New process technology development must be carried out to achieve higher frequency modules and higher levels of integration. The processes for forming wiring conductors and the processes for dielectric layer fabrication are described separately below.

(1) Conductor fabrication processes

With the miniaturization of components, wiring dimensions are becoming progressively finer and technology development directed to forming fine conductors is being emphasized. In the miniature filter application described above, the conductor wiring formed in the material surface must have a high dielectric constant as well as ultra fine line width. If fine wiring cannot be formed, it will not be easy to use materials with a high dielectric constant in devices.

With the current thick film printing processes, line widths of 80 μm are the limit. The development of new processes is desirable for achieving finer wiring. The reason for the difficulties in achieving miniaturization lies in the principles of the screen printing technique itself in which ink is pressed through openings in the mask, although the surface roughness of the substrate also has a significant impact. For example, below is one of the solution processes. First, conductor wiring is screen printed on plastic film with a smooth surface. Then the printed surface is applied to the green sheet and a method of transfer is applied, so that line widths of around 40 μm can be achieved.

As indicated in the previous section, it is effective to reduce the surface roughness of the conductor in order to reduce conductor loss. For this reason it can be considered necessary to develop processes to print the conductor after flattening the cast green sheet by applying pressure, or processes to apply pressure to the conductor after printing to make the conductor wiring flat and so on.

(2) Dielectric fabrication processes

In order to develop modules using materials appropriate to a variety of functions, high level process technologies to incorporate ceramics with different characteristics such as dielectric constant will be necessary.

To combine dissimilar materials while achieving good consistency, technology development of firing processes in particular will be necessary [8]. The following three problems with technologies for combining dissimilar materials can be identified.

- Mismatches of firing shrinkage behavior between dissimilar materials
- Interactions between dissimilar materials
- Mismatches of thermal expansion behavior between dissimilar materials

The first point, the mismatch of firing shrinkage rates, is the biggest problem for combining dissimilar materials. To solve this, the "non-shrinkage process" has been proposed in which firing shrinkage in the x and y axes is suppressed, allowing shrinkage only in the z axis (Ref. Chapter 8) [9, 10].

The second point, interaction, in the case of LTCCs occurs frequently through the glass phase in the ingredients. Therefore it is necessary to examine the reactions between the glass in the respective materials. For example, with crystallized glass, the composition of the glass changes as a result of reaction with amorphous glass so that it is difficult to precipitate a specific crystal phase, and the resulting dielectric characteristics may vary markedly.

Furthermore, when the finished fired substrate is heated and cooled repeatedly, if the mismatch between the thermal expansion of the dissimilar materials is significant, it has been pointed out that cracking and so on may occur between the dissimilar materials.

In order to achieve finer wiring patterns as described above to enable the miniaturization of components, since the margins for alignment precision will get smaller, there is a requirement for greater exactness in quality control in matters such as dimensional change in the green sheets and the like.

To raise the level of the technology, it is likely that besides improving LTCC technology, it will also be necessary to combine other materials and other technologies to establish new techniques.

10.3 Background of post-LTCC technology

As described above, development of small high frequency devices and circuit boards with a variety of integrated passive components has been proceeding at a rapid pace, while the requirements and specifications for the electrical characteristics, size and cost of the devices have been getting ever more demanding.

As shown in Table 10-1, the development of devices with integrated high frequency passive components currently involves three techniques, including LTCCs.

Resin printed circuit board technology is a technique in which multiple layers are built up through laminating epoxy resin layers on FR4 substrates. Conductor wiring is formed by plating with copper, while the via holes between epoxy layers are made using a laser [11]. With this technique, line widths of 50 μm are the limit. When embedding or integrating capacitors and the like in the structure, the material used for the capacitors is a ceramic/polymer composite consisting of an epoxy resin base with ceramic powder dispersed in it. Since achieving a high dielectric constant with the composites is difficult, it is hard to make embedded capacitors with large capacity, and the capacity density that can be achieved is in the region of several dozen pF/cm^2. Since the materials and processes are inexpensive, low cost production is possible. However, as epoxy based resin is used as the dielectric material, the high frequency characteristics of the structure are not good (dielectric loss at high frequency is high), and it is not suited to high frequency applications.

In the process using the MCM-D application of silicon technology where multiple layers of thin film are formed on a silicon wafer, vacuum processes such as sputtering and the like are used for the conductor wiring, while resins like polyimide and so on are used for the interlayer insulating layer [12, 13]. Fine wiring of less than 10 μm can be achieved. $BaSrTiO_3$ thin film (film thickness: several hundred nm) with a dielectric constant of around 400 formed by sputtering or the sol-gel process can be applied to capacitor materials. However since it is necessary to anneal the substrate in an atmosphere containing oxygen after forming the film to improve the dielectric constant, it is difficult to apply it to internal functions with a wiring system that uses copper. If this problem can be overcome, it will be possible to achieve a capacity density of some several hundred nF/cm^2 due to the thinness of the film in the dielectric layer. Because this method requires use of photolithography processes using vacuum equipment in a clean room, it costs more than the other methods. In addition, the polyimide resin used in the interlayer insulation has very good high frequency characteristics for a resinous material, although since it is inferior compared to ceramic material, it is hoped that its characteristics can be improved.

Table 10-1   A comparison of the state of the art of various kinds of process.

Requirement	Small size, fineness	Integration	Cost	High frequency	Process
Current characteristics	Smallest Wiring line width (μm)	Passive components Embeddability [Embedded capacitor capacity]		Dielectric loss (tan δ)	
Printed circuit board technology	50	Several dozen pF/cm^2	Low	×	Plating, lamination
Silicon technology (MCM-D)	5	Several hundred pF/cm^2	High	△	Sputtering (vacuum) Photo-lithography
LTCC	50	Several dozen nF/cm^2	Average	O	Screen printing, high temperature firing

× - Does not meet requirements, △ - Partially meets requirements, O - Meets requirements

As noted in Chapter 1, LTCCs are made by printing the thick film conductor wiring on ceramic green sheets, and after the green sheets are laminated, they are cofired at a high temperature of around 1,000°C. Since the thick film printing process is used, line widths of 50 μm are the limit. Composites of high dielectric constant ceramic and glass are used as materials for embedded capacitors, and as these materials are formed in sheets inside the multilayered structure, a capacity density of several dozen nF/cm^2 can be formed. Because the firing process uses high temperatures, there are limits to cost reduction although compared with the silicon technology mentioned above, a lower process cost can be achieved. As the high frequency characteristics of ceramic material are superior to that of resins, they are better suited to high frequency applications.

As shown in Table 10-1, since compared with the other technologies LTCCs meet the requirements more fully, they are currently the most promising technology for high frequency integrated modules. However, as can be judged from the table, they do not fulfill all the requirements. It is hoped that in the future modules will meet all the requirements. In order to realize this, technology development that fulfills the following four requirements for materials and processes must be pursued; 1) Photolithography processes from the point of view of small size and fineness, 2) use of low cost resin FR4 boards as base substrates, 3) low cost plating process technology for wire formation, and 4) ceramic materials with excellent high frequency characteristics. Figure 10-4 shows a module which incorporates the requirements above. Ceramic material with a high dielectric constant, and low loss materials appropriate to the required high frequency functions, are incorporated into epoxy resin sheets that are built up in several layers on an FR4 substrate. Copper plating technology is used for the wiring [14].

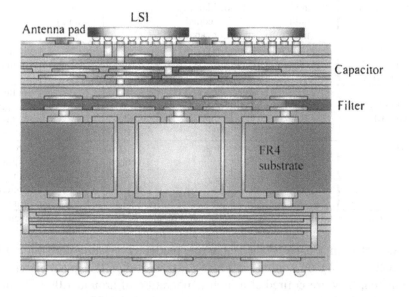

Figure 10-4   View of a future high frequency module.

The technology development that is the key to realizing high frequency modules like this is ceramic film deposition technology applied to resin material. The following three qualities are required of ceramic film.

(1) Deposition at low temperatures below the level of heat resistance of resins (around 250°C in the case of epoxy resin).
(2) Excellent dielectric characteristics close to that of the bulk material (for example, dielectric constant of 1,000 or more).
(3) Deposition of a film thickness appropriate to the surface roughness of the build-up substrate.

Table 10-2 is a comparison of various kinds of ceramic deposition technologies. When forming a ceramic film with the sputtering method, deposition on a resin substrate can be considered difficult since post-annealing at 300°C at least is required. However, when post-annealing at around 600°C is performed, a dielectric constant of around 500 can be obtained [15]. In addition, deposition at the micron level is difficult with this method. As with the sputtering method, post-annealing at 300°C is also required with the sol-gel process, so that deposition on a resin substrate is difficult. The dielectric constant that can be obtained is lower compared with the sputtering method and after post-annealing, a value of around 400 is the limit [16]. As for film thickness, a thickness of around 5 μm can be obtained through multilayer coating. With thick film processing the film is deposited by screen printing a paste made of ceramic powder on the substrate then firing it at a high temperature of around 1,000°C. Although a thick film with dielectric characteristics close to that of the bulk material can be obtained, the processing temperature is high so that it cannot be applied to resin substrates. Composite films of ceramic and resin can be obtained by evenly combining ceramic powder with a high dielectric constant such as $BaTiO_3$ and resinous varnish compound and coating it on the substrate, then heat curing it at around 200°C. Although the requirements for processing temperature and film thickness are met, a high dielectric constant cannot be achieved. Georgia Institute of Technology reported that they obtained a composite with a dielectric constant of around 135 to 150 by first treating the surface of the ceramic particles and by optimizing ceramic particle diameter and the composition of the resin compound, then introducing mixed powder of $Pb(Mg_{1/3}Nb_{2/3})O_3$-$PbTiO_3$ ceramic (dielectric constant: around 15,000) and $BaTiO_3$ (dielectric constant: around 14,000) at 85 vol% to epoxy resin with a dielectric constant of 3.2 [17, 18, 19, 20]. With this method of mixing and compositing high dielectric constant ceramic and resin, the dielectric constant of around 150 reported by the Institute is probably the limit. However the aerosol deposition (AD) method offers the possibility of meeting all three requirements above.

10.3.1 AD process as a post-LTCC technology

Aerosol deposition is a revolutionary film production technology for making ceramic films at room temperature, devised by Dr. Jun Akedo of the National Institute of Advanced Industrial Science and Technology [21, 22]. Figure 10-5 is an outline of the AD equipment. An aerosol flow (a mixture of fine ceramic particles and gas) is created by supplying compressed gas to agitated raw material dry ceramic fine powder. The flow is further accelerated in a vacuumed pressure atmosphere (around 50 to 1 kPa) using a vacuum pump, and passing through a slit-shaped nozzle, it is sprayed onto the substrate as an aerosol, forming a ceramic film. The ceramic particles of the raw material with a particle size of 0.05 to 2 μm are accelerated to between 100 to 1,000 m/sec. and the ceramic produces a film on the substrate at a rate of about 10 to 30 μm/min at room temperature.

Since no rise in temperature is observed in the vicinity of the substrate, film production on the surface of resinous materials such as plastic is possible. As the fine raw material particles are not broken down to the molecular level in the film production process, even multicomponent complex compounds do not undergo composition shifts, so that the method has the merit of being able to produce films with a complicated composition. So far, this method has been used to produce films of PZT piezoelectric ceramics and alumina.

Table 10-2   A comparison of various kinds of ceramic deposition technologies.

Requirement	Low processing temperature	High dielectric constant	Thick film
	Up to 200°C	1,000 up	1 to 10 μm
Sputtering method	△ (300°C up)	△ (around 500)	×
Sol-gel process	△ (300°C up)	×	△ (up to 5 μm)
Thick film processing	×	○	△ (5 μm up)
Ceramic/resin composite film	○	×	△ (5 μm up)
Aerosol deposition (AD) method	○	○	○

× - Does not meet requirements, △ - Partially meets requirements, ○ - Meets requirements

10.3.2   Current and future state of development of AD ceramic film

Our group within Fujitsu was largely responsible for the development of film production technologies and device manufacturing technologies using high frequency, high dielectric constant dielectrics represented by $BaTiO_3$ as part of the Ministry of Economy, Trade and Industry's Nano Structure Forming for Advanced Ceramic Integration Technology Project which began in 2002 [23, 24, 25].

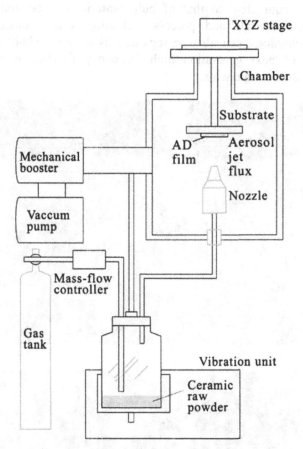

Figure 10-5   An outline of aerosol deposition AD equipment.

Figure 10-6 (a) is a cross sectional photograph of the BaTiO$_3$ composition of AD film formed on a resin substrate. A dense structure packed with ceramic particles of several dozen nm can be observed. The dielectric characteristics (@ 100 kHz) that can currently be obtained are dielectric constant: 400, tan δ 2% or so (dielectric constant of bulk BaTiO$_3$: around 3,000). Figure 10-6 (b) is the microstructure of the microwave dielectric structure of Ba(Zn$_{1/3}$Ta$_{2/3}$)O$_3$-Al$_2$O$_3$ AD film. This exhibits a microstructure of alumina particles distributed uniformly in a BZT matrix. The Q value (1/tan δ) of the film is 500 (Q value of bulk Ba(Zn$_{1/3}$Ta$_{2/3}$)O$_3$: around 5,000).

Since the AD method uses ceramic powder as described above to form a film, knowledge taken from the ceramics industry and ceramic materials engineering can be used to good effect. The development of dielectric film deposited with the AD method started only recently, but with further technology development, the fabrication of ceramic film with dielectric characteristics equivalent to that of bulk material can be anticipated as various other materials and process technologies are considered. AD deposition technology can also be regarded as a core technology for the development of next generation high frequency function modules with integrated passive components.

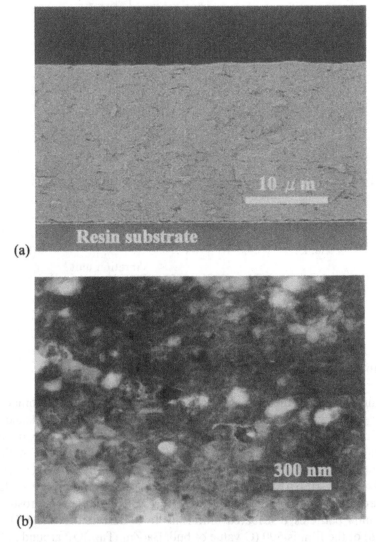

(a)

(b)

Figure 10-6   (a) Cross sectional structure of the $BaTiO_3$ composition of AD film formed on a plastic substrate, (b) microstructure of the microwave dielectric structure of $Ba(Zn_{1/3}Ta_{2/3})O_3$-Al_2O_3 AD film, white spot: alumina particle.

Acknowledgement

The research of this chapter was partially supported by NEDO projects of "Nano Structure Formation for Advanced Ceramic Integration Technology in Japan – the nano technology program".

References

[1] "Development of Ubiquitous Service using Wireless Technology", NTT Technical Journal, No. 3 (2003) pp. 6-12.

[2] "Restructuring System on a Chip Strategy with Package Technology as the New Innovation", NIKKEI MICRODEVICES, No. 189 March (2001) pp. 113-132.

[3] "Activity Around Technology to Embed Devices Internally in PCB's Suddenly Increases", NIKKEI ELECTRONICS, No. 842, March 3 (2003) pp. 57-64.

[4] Y. Imanaka, "Material Technology of LTCC for High Frequency Application", Material Integration, Vol. 15, No. 12, (2002), pp. 44-48.

[5] A. A. Mohammed, "LTCC for High-Power RF Application?", Advanced Packaging, Oct. (1999), pp. 46-50.

[6] H. Sobol and M. Caulton, Advances in Microwaves, No. 8, (1994), pp. 11-66.

[7] Y. Imanaka and M. R. Notis, "Metallization of High Thermal Conductivity Materials", MRS Bull., June (2001), pp. 471-476.

[8] Y. Imanaka and N. Kamehara, "Influence of Shrinkage Mismatch between Copper and Ceramics on Dimensional Control of Multilayer Ceramic Circuit Board", J. Ceram. Soc. Jpn., Vol. 100, No. 4, (1992), pp. 560-564.

[9] H. T. Sawhill, R. H. Jensen, K. R. Mikeska, "Dimensional Control in Low Temperature Co-fired Ceramic Multilayers", Ceramic Transactions, Vol. 15, (1990), pp. 611-628.

[10] K. R. Mikeska, and R. C. Mason, "Pressure Assisted Sintering of Multilayer Packages", Ceramic Transactions, Vol. 15, (1990), pp. 629-650.

[11] T. Nishii, S. Nakamura, T. Takenaka, and S. Nakatani, "Performance of Any Layer IVH Structure Multi-layered Printed Wiring Board", *Proc 18th Japan International Electronic Manufacturing Technology Symposium (IEMT), Omiya, Dec. 1995*, pp. 93-96

[12] H. Yamamoto, A. Fujisaki, and S. Kikuchi, "MCM and Bare Chips Technology for Wide Range of Computers", *Proc 46th Electronic Components and Technology Conf*, Orlando, FL, May. 1996, pp. 113-138.

[13] K. Prasad, and E. D. Perfecto, "Multilevel Thin Film Applications and Processes for High End System", *IEEE Trans-CPMT-B*, Vol. 17, No. 1 (1994), pp. 38-49.

[14] Y. Imanaka, "Technology for obtaining high capacitance density in substrate integrating passive function", Embedding Technology of Passive Component in Printed Wiring Board, Technical Information Institute, (2003), pp. 154-161.

[15] S. Yamamishi, H. Yabuta, T. Sakuma, and Y. Miyasaka, sputtering, "(Ba+Sr)/Ti ratio dependence of the dielectric properties for $(Ba_{0.5}Sr_{0.5})TiO_3$ thin films prepared by ion beam sputtering", Appl. Phys. Lett., Vol. 64, No. 13 28 March (1994), pp. 1644-1646.

[16] Y. Imanaka, T. Shioga, and J. D. Baniecki, "Decoupling Capacitor with Low Inductance for High-Frequency Digital Applications," *FUJITSU Sci. Tech. J.*, Vol. 38, No. 1 June (2002), pp. 22-30.

[17] P. Chahal, R. R. Tummala, M. G. Allen, and M. Swaminathan, "A Novel Integrated Decoupling Capacitor for MCM-L Technology", Proc.46th Electronic Components and Technology Conf., Orland, FL, May(1996) pp. 125-132.

[18] V. Agarwal, P. Chahal, R. R. Tummala, and M. G. Allen, "Improvements and Recent Advances in Nanocomposite Capacitors Using a Colloidal Technique", Proc. 48th Electronic Components and Technology Conference, Seatle, WA (1998), pp. 165-170

[19] S. Ogitani, S. A. Bidstrup-Allen, and P. Kohl, "An Investigation of Fundamental Factors Influencing the Permittivity of Composite for Embeded Capacitor", Proc. 49th Electronic Components and Technology Conference, San Diego, CA (1999), pp. 77-81.

[20] H. Windlass, P. M. Raj, S. K. Bhattacharya, and R. R. Tummala, "Processing of Polymer-Ceramic Nanocomposites for System-on-Package Application", Proc.51st Electronic Components and Technology Conf., Orland FL, May(2001) pp. 1201-1206.

[21] J. Akedo and M. Lebedev, "Piezoelectric properties and poling effect of Pb(Zr, Ti)O_3 thick films prepared for microactuators by aerosol deposition", Applied. Physics. Letter, Vol. 77, No. 11 (2000), pp. 1710-1712.

[22] J. Akedo, and M. Lebedev, "Ceramics Coating Technology Based on Impact Adhesion Phenomenon with Ultrafine Particles-Aerosol Deposition Method for High Speed Coating at Low Temperature-", Materia Japan, Vol. 41, No. 7 (2002) pp. 459-466.

[23] Y. Imanaka, "Material Technology of LTCC for High Frequency Application", Material Integration, Vol. 15, No. 12 (2002) 44-48.

[24] Y. Imanaka, J. Akedo, "Integrated RF Module Produced By Aerosol Deposition Method", Proc 54th Electronic Components and Technology Conf, Las Vegas, NV, June. (2004), pp. 1614-21 .

[25] Y. Imanaka, J. Akedo, "Passive Integration Technology for Microwave Application Using Aero-Sol Deposition", Bull. Ceram. Soc. Jpn., Vol. 39, No. 8, (2004), pp. 584-589.

Index